The Human Body:
An Intelligent Design

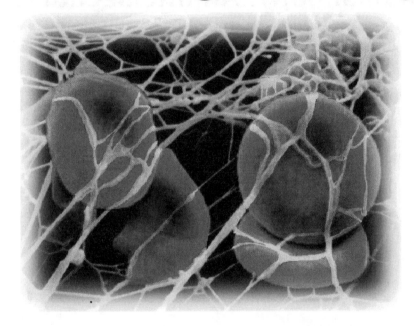

Alan L. Gillen   Frank J. Sherwin III   Alan Knowles

Editor: George F. Howe   Illustrator: Miriam Rodriguez

Creation Research Society Monograph Series: No. 8

Creation Research Society Books

The Human Body: An Intelligent Design

Alan L. Gillen, M. S., Ed. D.
Frank J. Sherwin III, M. A.
Alan Knowles, M. S.

ISBN 0-940384-21-3

Copyright 2001 Creation Research Society
SECOND EDITION

Printed in the United States of America

All scripture quotations taken from the King James (Authorized) Version of the Bible.

Cover depicts a scanning electron micrograph of blood clotting.
(Photograph modified from Davis & Kenyon, 1993, with permission)

Foreword

A Needed Approach to the Study of the Human Body

If there is no Creator, there is no real purpose to life. If there is no designer, there is no design but just a network of unrelated, non-functioning, loosely assembled parts. When a student first examines the human body, he is overwhelmed by the complex and diverse designs that give precise functional purpose to each structure. Further study and patient research reveal repeating physiological patterns in the eleven body systems. This book describes these patterns and discusses their purpose and meaning. It will help the teacher/serious student of biology to appreciate the creative design principles and organization of the human body.

This book discusses the various body systems by comparing and contrasting the viewpoints of intelligent design versus a blend of matter using random processes over vast periods of time with mutations resulting in the phenomenon called natural selection. It teaches about the biological basis of blood clotting, the remarkable immune response, recent research on split brain studies, the physiology of flight, the body's adaptation to high altitudes and concludes that the human body is the "ultimate machine."

This book is built around the basic universal patterns and themes in human biology that include 1) the direct relationship of structure to function, 2) the role of programmed homeostasis for precise functioning of metabolic mechanisms, 3) the interdependence between body parts, 4) short-term physiological adaptation, 5) maintenance of membranes/boundaries and 6) the triple concepts of order, organization and integration. Most popular books on the human body, such as best-selling National Geographic titles, *The Incredible Machine* and *Incredible Voyage: Exploring the Human Body*, as well as most human anatomy and physiology texts, assume an evolutionary development of matter, life and the human body. This book is unique in that it is built around the widely accepted structural and functional themes, but provides a distinct creationist approach to the study of the human body. It is ideal for students of basic biology, human anatomy and physiology, pre-medical studies and all others interested in digging deeper into the logic that the exquisite design in the human body infers a Divine Designer. It challenges biology students to evaluate whether the creation or evolution model of origins makes more sense.

This book is a must, either as a basic text or as a supplement to a basic text, for all human anatomy and physiology courses in Christian colleges and universities.

David A. Kaufmann, Ph.D., F.A.C.S.M.
Professor (Retired), Department of
*   Exercise and Sports Science*
University of Florida
Gainesville, Florida
August 25, 1998

Preface

We are living in an exciting time when technology makes possible more observations in the human body than at any time before. We have the intelligence to propose sophisticated explanations for what we see. At first study, complexity and diversity overwhelms the mind when we examine the human body. Upon continuous study and patient research, repeating physiological patterns emerge in the eleven body systems. We want to describe these patterns and discuss their meaning.

We wrote *The Human Body: An Intelligent Design* to help readers: 1) understand the basic universal themes in human anatomy and physiology; 2) compare and contrast two viewpoints of man's origin; and 3) illustrate and apply physiological principles in the human body from the perspective of intelligent design. Despite all society has learned, answers given for the big questions remain uncertain. Uncertainty keeps science alive. Our purpose is to help readers understand the human body from a biblical perspective. There is no attempt to mislead you to tell you that a complex issue is simple, or that the authors' views are the only reasonable ones.

Here we give a favorable case for creation, or intelligent design, and we raise reasonable doubt about macroevolution theories. The majority of college level biology and physiology textbooks bring an evolutionary perspective. But you have a mind of your own and we hope you will compare the views presented here with contrasting views of human anatomy and physiology

in secular books. We hope that you will enjoy dealing with the ideas and dig deeper into each subject. We expect that reading *The Human Body: An Intelligent Design* will become an exciting event in your educational journey.

This book has also been written to fill a gap in the literature on evidences of design found in nature. Very few books and articles address a creation perspective on human anatomy and physiology. Five notable exceptions in book form include *Our Amazing Circulatory System: By Chance or Creation* (Clark, 1976), *Fearfully and Wonderfully Made* (Brand and Yancey, 1980), *In His Image* (Brand and Yancey, 1984), *The Amazing Body Human* (Cosgrove, 1987), and *The Human Body: Accident or Design?* (Jackson, 1993). These books are either out of print, however, or have become difficult to obtain. Although their principles are timeless, each of these books is somewhat dated in its detail. Technology and knowledge have advanced over the past decade and we feel a need to provide an updated book on design of the human body.

The periodical literature, other than the *Creation Research Society Quarterly*, also lacks discussion on this most important topic. One of the few active researchers and writers in the discipline of human anatomy and physiology from a creation perspective is Dr. David Kaufmann, Professor of Exercise Science, University of Florida. Dr. Kaufmann, an accomplished researcher in exercise physiology and a board member of the *Creation Research Society*, made numerous suggestions to the earliest outline of this book. He has generously donated material for this book from his professional writings in physiology.

The authors want you to use this book as a supplement, not as a substitute for your anatomy and physiology textbook. But without this supplement, you will miss a lot of interesting science. We hope when you finish this book, you better understand the human body from a traditional, biblical perspective. The subjects are treated in-depth and digging deeper brings richer rewards. Your textbook provides an extensive treatment of anatomy and physiology on each system, as well as a broader range of topics. Alternately consult both books, using each to enrich the other.

Finally, we acknowledge a number of people who contributed in one way or another in writing this book. Without them, this book may have not been completed in this form. We are grateful for another *Creation Research Society* board member and former editor of the *Creation Research Society Quarterly*, Dr. George Howe, who edited this book. We are extremely grateful for his suggestions, especially in his reading, rewriting, and proofing the many drafts for this book manuscript. In addition to his editorial skills, he provided technical expertise in biology. We gratefully acknowledge Dr. Joel Porcher, Assistant Dean of Arts and Sciences, for his editorial help in the first edition which piloted at Pensacola Christian College (PCC). His suggestions, along with PCC faculty, staff and students, help produce a better book. In particular, we acknowledge Becky Knowles, Jayne Gillen, and Dr. Carlos Alvarez for their proofing various drafts of this book. Also, we thank Miriam Rodriguez who contributed several excellent figures for the book. While we acknowledge the assistance of these individuals, we alone are responsible for the content and for the viewpoints expressed.

Alan L. Gillen
Pensacola Christian College

Frank J. Sherwin III
Institute for Creation Research

Alan Knowles
Pensacola Christian College

Dedication

This book is dedicated to our families who have lovingly supported us during the time of our research and writing about God's amazing creation, the human body. In addition, we wanted to acknowledge our students at Pensacola Christian College who first heard these lectures on the design principles of anatomy and physiology and who encouraged us to keep presenting truth in a world that seeks only what is popular.

Who hath put wisdom in the inward parts?
Or who hath given understanding to the heart?
Job 38:36

Table of Contents

Chapter One
Design Principles of the Human Body

We are indeed a splendid invention, capable of learning, insatiably curious about ourselves and the rest of nature, anxious to be loved, hoping to be useful.

Lewis Thomas, M.D. (1986, p. 7)

An artistic snowflake, a beautiful butterfly, a *Coleus* leaf, a panda bear, the complex vertebrate eye, and the hand of a surgeon are all examples of intricate designs found in nature. The origin of these patterns is a topic that fascinates most biologists. A dictionary definition of *design* is a plan, a scheme, a project, or a purpose with intention or aim. Many of these natural patterns and body plans in animals have been described as having a "perfect fit" with their environment. Are these observed designs the product of chance and natural selection, or are they the "fingerprints" of a master creator?

Many biologists view man as the product of cosmic evolution from some hominid ancestor. Still other biologists question this naturalistic model of human descent because there is a unique plan and pattern to the human body. Today, many biologists are reconsidering design and are seeing *Homo sapiens* as the pinnacle of design because of his spectacular cell biology, anatomy and physiology.

The purpose of this book is to: 1) describe basic universal themes in human anatomy and physiology; 2) compare and contrast the viewpoint of intelligent design with the neo-Darwinian explanation of anatomy;

and 3) to illustrate and apply physiological principles in the human body from a creation perspective. Many of the terms that are useful in discussing such a plan can be found in the glossary (Appendix A).

The Human Body and Its Design

What a piece of work is a man how noble in reason! how infinite in faculties! in form and moving how express and admirable in action like an angel, in apprehension how like a god!

Shakespeare
(Brand and Yancey, 1984, p. 16)

Awesome, incredible, or ingenious are some of the adjectives that men through the ages have used to describe the order found in the human body. The splendor of the human body can only be described in superlative terms! When one considers the movement in the hand of a concert pianist, the thought processes in the brain of a heart surgeon, the eye focus required of a seamstress, and the muscle coordination that propels a world class gymnast, it is difficult to imagine this body plan has happened by chance. A naturalistic explanation alone cannot account for the incredible complexity and optimal integration in human anatomy and physiology. All these life processes require incredible movement, coordination and communication among the body's organs and cells.

The human body consists of eleven organ systems, four basic body tissues, and dozens of different specialized cells (Marieb, 1994). The human body is mostly made up of an estimated 30 to 100 trillion cells, with most estimates counting over 75 trillion cells. This is quite a range! Perhaps this range is so vast because of the diversity of human sizes from Billy Barney, a famous circus midget, to Hakeem Olajuwon and Shaquille O'Neil (NBA stars). Most of these cells can be seen with a light microscope. Some 100 million are red blood cells, and several hundred million are nerve cells. The human body is truly a highly organized and coordinated system!

Human Body as a Machine

The human body is like a machine in many ways. The body and a machine both perform work. This analogy is not new. It was used by Renaissance scholars, including the famous artist and scientist, Leonardo da Vinci. At the close of the 15th Century, Leonardo da Vinci made the most comprehensive study of the human body, yet, he saw neither superfluous nor defective structure in man. In fact, he described human anatomy as one of beauty and complexity. In addition, he made sketches of the body, in a study of proportions, and compared them with the most sophisticated machines in his time. Because the body was so masterfully engineered like a "machine," it has been the subject of many artists' work through the centuries.

This machine analogy is still applicable today. Each part of the body has its own job. The parts work together to keep the body alive, much as the parts of an automobile work together to make it run. The skin, for example, protects the body as paint protects the metal on a car. Food serves as fuel for the body as gasoline powers a car engine. The human body wears out and breaks down if not properly maintained. If a machine requires a blueprint or architect, how much more does the plan of the human body suggest that it has a Maker?

Remember, however, that the human body is not as simple as a machine made by people. If a car breaks down, its broken parts can be replaced. If some parts of the body wear out and break down completely, they cannot so easily be replaced to make the body as good as new. Many parts such as the hair and outer layers of skin, however, are continually being replaced as older portions die and fall off. Unlike a machine, the body can heal itself within limits as illustrated in the case of a broken bone that forms a bone collar and a wound that disappears as the tissue is restored.

Basic Themes of Human Anatomy and Physiology

There are basic themes, or principles, that can be observed in all eleven of the human body systems. These themes include the 1) relationship of structure to function; 2) steady state of metabolism, or homeostasis; 3) interdependence among body parts; 4) short-term physiological adaptation; 5) maintenance of boundaries; and 6) the triple scheme of order, organization, and integration. These themes are widely discussed by physiologists and are consistent with a creation perspective of the human body.

The **correlation of structure and function** can be explained by stating that, in general, the physical form of an animal tissue, organ, or system is related to its function. Two examples of this are 1) the composition of bone that makes it both strong and relatively light to handle the body's weight, and 2) the longitudinal separation in the heart that keeps unoxygenated from oxygenated blood. Both are each exquisite demonstrations of form being related to function.

The human body maintains itself in equilibrium, a steady state known as **homeostasis**. This concept is based upon feedback that prevents small changes from becoming too large and harmful. Changes occur between internal and external environments, and between interstitial fluid (the fluid between our cells) and intracellular fluid. "Thermostat wars" inside the house frequently happen when family members are at home; one member is hot and the other is cold. Clearly, the body's hypothalamus or body "thermostat" has a unique best setting. The body will regulate its temperature by a negative feedback mechanism by either shivering to warm the body when it is cold or sweating to cool the body when it is hot.

Another example of homeostasis and negative feedback is glucose regulation by insulin and glucagon. The pancreas is considered both an exocrine and endocrine gland. The endocrine function of the pancreas secreting insulin in the blood is controlled by the amount of glucose in the blood. The pancreatic cells that control blood glucose levels are called Islets of Langerhans. Insulin and glucagon work as a check and balance system regulating the body's blood glucose level. Glucagon accelerates the breakdown of glycogen to glucose in the liver (glycogenolysis) and this causes an increase of blood sugar. Insulin is antagonistic to glucagon. It decreases the blood glucose concentration by accelerating its movement out of the blood and through the cell membranes of the working cells.

As glucose enters the cells at a faster rate, the cells increase their metabolism of glucose. All sugary and starchy foods, such as bread, potatoes, and cakes are broken down into glucose. In this form they can be absorbed by every cell in the body, including the cells in the liver, which store glucose in the form of glycogen. Cells absorb glucose and create ATP using the energy released from the sugar.

ATP is generated in the cell's mitochondria, resulting in an energy storage and the production of carbon dioxide and water as byproducts. This aerobic process is the body's principle source of energy and it cannot take place without insulin. One type of diabetes occurs when the pancreas fails to secrete enough insulin and so fails to regulate the glucose concentration in the blood. The normal glucose level for an average adult is about 80 to 120 milligrams of glucose in every 100 milliliters of blood. If the beta cells of the pancreas secrete too little insulin, an excess of glucose collects in the bloodstream causing *diabetes mellitus* (hyperglycemia), the most common disorder of the endocrine system.

Many early anatomists believed that body parts share a common thread uniting them; they are **interdependent** with each other. These early workers noted that the body is one unit, though it is made up of many different types of cells and tissues. They all form one body as noted by Brand and Yancey, 1980. Related terms that other scientists have used to describe this phenomenon of interdependence include an "adaptational package", "cell team", "compound traits", "emergent properties", "irreducible complexity", "molecular team" and "synergism" (Campbell, 1996; Davis and Kenyon, 1993; Parker, 1996). This resulting condition of interdependent body parts working together is synergistic such that the sum of their action is greater than the addition of separate, individual actions.

Two examples of the marvelous advantage resulting from this cooperation at all levels from the molecular to the systematic are 1) the amazing interdependence of many parts for focusing the lens in the human eye; and 2) the irreducible complexity involved in the cascade (meaning one effect or adjustment must occur in order for the next level of adjustment to happen) of biochemical reactions required for blood clotting.

Adaptation allows living cells to adjust the body to change in the external environment. Short-term change is allowed by physiological adaption, an example of which is the changing oxygen levels in our bloodstream at different altitudes. Oxygen pressure in the atmosphere decreases at higher altitudes. The athlete who is conditioned at sea level will have trouble breathing at high altitude stadiums. If the body remains living and training at higher altitudes over a period of months, however, then it will physiologically adapt to this altitude by increasing the level of a hor-

mone called erythropoietin. This hormone increases the number of red blood cells, and in turn, the oxygen available to body tissues also increases (Guyton, 1991). This adaptation gives greater "wind" to the athlete competing in "mountain" arenas.

Every living organism must be able to **maintain its boundaries** so that its inside structures remain distinct from its outside chemical environments. Every human body cell is surrounded by a cell membrane that encases its contents and allows needed substances to enter while restricting the entry of potentially damaging or unnecessary substances. Additionally, the whole body is enclosed by the integumentary system, or skin. The integumentary system protects internal organs from drying out, from bacterial invasion, and from the damaging effects of an unbelievable number of chemical substances and physical factors in the external environment.

Finally, the triple theme of **order**, **organization,** and **integration** can be clearly seen in many of the human body systems. These topics are also illustrated in the anatomy and physiology of other animals. The levels of organization from least complex to most complex are molecules, cells, tissues, organs, organ systems, and organisms. A plan and purpose can be seen through the structure and function of an information system, leading us to believe the parts of the animal and the human body are the work of an intelligent designer. This argument may apply from the molecular level to the gross anatomical level.

Other factors that affect the human body plan include its **bioenergetics**. Bioenergetics is fundamental to all animal functions. Man derives chemical energy from his environment through various foods, as well as various micro and macronutrients that are assimilated from these organic molecules. Eventually, the energy of these nutrients is used to synthesize **ATP** to meet the energy needs of the organ systems.

An Anatomical Anomaly

Scientists must be cautious not to get too set in their ways. Good scientists are open to alternative explanations as new data become available. This is true even in the old discipline of gross anatomy. *Gray's Anatomy* is the "Bible" of human musculature, but even this "Bible" may be incomplete. In 1996, Dr. Gary Hack, a research dentist, announced that he and his coworkers had found a new muscle in the face (Dunn, Hack, Robbins, and Koritzer, 1996). Hack stumbled onto the structure while working on a cadaver so altered by previous dissections that he was forced to cut into the face from the front instead of the usual side approach. Hack found this new muscle, the **sphenomandibularis,** by cutting from an unconventional angle and exposing an unfamiliar muscle connecting the mandible (lower jawbone) to the sphenoid bone behind the base of the eye socket. He suspected that it stabilizes the jaw during chewing rather than actually moving the jaw.

Hack still needs evidence to show that this new structure is indeed a separate muscle. In order for it to be considered a separate muscle, the sphenomandibularis must have its own nerve and blood supply, connect at both ends of the skeletal structure, and have a distinct function. But no matter what future findings may be, it is significant that no description of this structure is discussed in *Gray's Anatomy* or any other of the fifteen "standard" anatomy textbooks. You might say that this finding has given researchers something to "chew on."

Hack's research has now caused the classic reference, *Gray's Anatomy* to be revised. The moral of this story is that a study of an old subject with alternative methods and an open mind may reveal new information and insight into how the body works. His work illustrates not only the expanding knowledge of human anatomy, but also that scientists need to keep an open mind for new data that may contradict established preconceived ideas.

"Vestigial Organs" Are Fully Functional

One topic of interest to anatomists over the years has been rudimentary or **vestigial** organs. A vestigial structure is a body part that apparently has no function, and is presumed to have been useful in ancestral species. Like the anatomical anomaly story discussed in the previous section, alternative research techniques and new ways of studying old topics have shed fresh insight into the study of "vestigial" organs.

Medical research now confirms that over one hundred body parts once considered "vestigial" do have functions. Some examples of previously imagined "vestigial" organs include the adenoids, appendix, thymus, tonsils, and lymph nodes (Bergman and Howe, 1990). Extensive research studies have now shown that each of the above organs is a part of our impressive immune system.

Tonsils have long been overlooked as an important body defense. Anatomists recognize three types of tonsils, palatine, pharyngeal (adenoids), and lingual. For hundreds of years, tonsils were thought to be vestigial organs, optional organs to be disposed. Tonsils are unusually large groups of lymph nodules and diffuse lymphatic tissue located deep within the oral cavity and the nasopharnyx. They form a protective ring of lymphatic tissue around the openings between the nasal and oral cavities and the pharnyx. This ringed boundary helps to provide a protection against bacterial, viruses, and other potential pathogens in the nose and mouth.

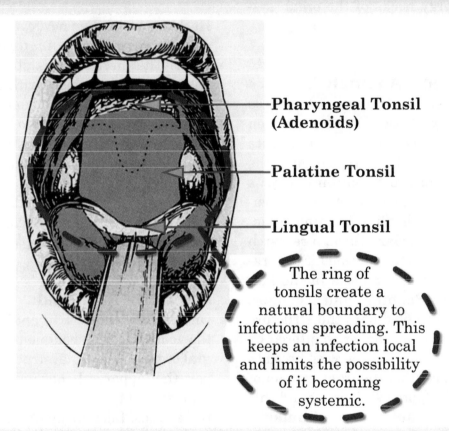

Pharyngeal Tonsil (Adenoids)

Palatine Tonsil

Lingual Tonsil

The ring of tonsils create a natural boundary to infections spreading. This keeps an infection local and limits the possibility of it becoming systemic.

In adults, the tonsils decrease in size and eventually disappear. Tonsils are often associated with the respiratory system because their location in the mouth and throat (therefore proximate to many respiratory organs). Due to their body location, there is a tendency to be infected by respiratory pathogens, such as *Streptococcus pyogenes* and adenoviruses.

Figure 1.1. Location of the Tonsils.

Many of the organs once classified as "vestigial" were found to be composed of lymphatic tissue, and hence, have an immune function. Two medically important organs routinely removed in former days were the tonsils (palatine tonsils) and the adenoids (nasopharyngeal tonsils). The long held assumption was that tonsils were useless

or vestigial, and hence tonsillectomies were frequent. Tonsillitis was over diagnosed (swollen tonsils during a cold), and the routine removal of tonsils by surgery was all too common. When one of the authors (ALG) was growing up, most physicians thought it best to remove these useless (vestigial) organs when the tonsils were irritated for more than a few days.

Tonsils were first suspected as a cause of health problems because of a connection between tonsil size and severity of respiratory infection. When children have the greatest severity of colds, tonsils swell. In the 1930s, over half of the children had their tonsils and adenoids removed. Then medical scientists learned that tonsils are important to young people in helping to establish the body's defense capabilities by producing antibodies. Once these defense mechanisms develop, the tonsils shrink to a smaller size in adults. Tonsils are important in the growth of the immunological system.

Tonsils are the first line of defense against respiratory viruses and bacteria. After learning these facts in 1971, Katz (1972) reported that *"if there are one million tonsillectomies done in the United States, there are 999,000 that don't need doing."*

Drawing Conclusions

Notice how flawed thinking yields false conclusions. Unfortunately, there were many needless surgeries performed in removing tonsils. We wonder how many malpractice suits might have been filed because of the error of removing fully functional organs. There is danger when scientists make errors in their conclusions. Many lives could have been spared. The number of severe colds and prolonged throat infections could have been reduced if physicians had not removed so many tonsils. In most cases, children would not have been exposed to the danger of this needless surgery. In addition, they would have had the benefit of a fully developed immune system to help fight respiratory diseases. In conclusion, the health of many individuals could have been improved if physicians had realized that tonsils and adenoids are fully functional.

Until very recently, vestigial organs were interpreted as strong evidence favoring macroevolution. For well over a century, the vestigial organ argument was considered one of the strongest supporting data of evolution and some writers still consider it to be a very strong support for evolution. But of the approximately 180 vestigial organs compiled by researchers around the year 1900, it is now almost unanimously agreed that most of these have at least one function in the body (Bergman and Howe, 1990).

In addition to the previously named organs of the immune system, the coccyx, plica semilunaris, and pineal gland were once considered examples of useless vestigial organs. Researchers have found that most of these so-called vestigial organs also play several roles. Some are backup organs, operative in unusual situations or during only certain stages of the organism's life. Such information has been very slow to find its way into the textbooks of biology and origins. For example, several major functions of the so-called nictitating membrane (plica semilunaris) in the human eye were delineated in the 1920's, yet some science text writers still label them "vestigial." Finally, we echo the conclusions of Bergman and Howe (1990, p. 85):

Since virtually all of the so called vestigial organs are shown to have functions, macroevolutionists can no longer credibly claim that evolution is the only origins model that will accommodate these scientific data. Individuals who are not fully indoctrinated with evolutionary philosophy will be able to see that all body organs function harmoniously. Dispelling the concept of vestigial organs allows the Creator's work in biology to be viewed scientifically as neither evolutionary, defective, nor capricious, but as evidence for His handiwork and design.

Chapter Two

Two Views of Body Design

When considering our origins it is clear that we have often been less than objective.

Richard Leakey (1982, p. 50)

There are two basic views regarding the origin of the universe, the earth, life, and man, the naturalistic view and intelligent design. The naturalistic view incorporates the belief in macroevolution (molecules to man), which has never been observed. It assumes that all life on earth is somehow related and that the origin and descent of all living things can be explained through a natural, mechanistic, random process. This concept, in the extreme, does not allow for the possibility of supernatural activity. The naturalistic idea has a long history that dates back to the ancient Greeks, but it became more popular in the 19th century. It found new support in speculations popularized by Charles Darwin in his book, *The Origin of Species*, published in 1859.

Darwin's Theory and Naturalistic Descent

The essence of Darwin's theory is that all life can be traced to a single ancestor through purely natural means. The plants, animals, and other organisms that surround us are products of random mutation acted upon by natural selection. This is commonly referred to as descent with modification. According to Darwin, nature acts like a breeder, scrutinizing every organism. When useful new traits appear, nature preserves them and passes them on to succeeding generations, while harmful traits are eliminated. Over time, these small changes accumulate until organisms develop new limbs, organs or other parts. Eventually organisms may change so drastically that they bear no resemblance to their original ancestor (Johnson, 1993; 1997).

Many evolutionists believe that all this happens with no purposeful input. In their view, chance, physiochemical laws, and nature run the whole show. In *The Origin of Species*, for example, Darwin (1979, p. 219) wrote:

If it could be demonstrated that any complex organ existed, which could not possibly have been formed by numerous, successive, slight modifications, my theory would absolutely break down. But I can find no such case.

In Darwin's time, however, no one appreciated the amazing complexity of living things. Back then, cells were thought to be little more than tiny blobs of gel. In recent years, Richard Dawkins (1996a), a biologist at Oxford University, in his book, *The Blind Watchmaker*, has popularized the neo-Darwinian theory. The essence of Richard Dawkins' argument is that given enough time (millions of years) and material (billions of individuals), many genetic changes will occur so that as a result slight improvements occur in structures such as the eye. Natural selection will favor these small improvements and they will spread though the human body over many generations. Little by little, one improvement at a time, the system becomes more and more complex, eventually resulting in a fully functioning, well-adapted organ. This does not mean that evolution can produce any conceivable structural change. For example, life forms do not have and probably will never have electronic gadgets. But evolution may be used as an explanation for complex structures, if we can imagine a series of small intermediate steps leading from the simple to the complex.

Because natural selection will act on every one of those intermediate steps, no single one can be justified on the basis of the final structure toward which it may be leading. Each step must stand on its own as an improvement that confers in itself an advantage on the organisms that possess it. Some changes will lead "upward" in overall evolutionary progress; still other changes, usually temporary ones, will go "downward." It is like a parking garage with ramps leading to different levels. You must travel the ramp to ascend or descend. Sometimes evolution goes three steps forward and two steps backward (like the so-called imperfections in life forms), but the overall progress is still forward.

Two views of origins are summarized in Table 2.1, "Predictions About the Nature of the Human Body." Although neither the naturalistic view nor the intelligent design view can be proven scientifically in the ultimate sense, each theory has predictions that may be tested. This table com-

pares each view's predictions about the human body's origins, genetics, development, organs, and systems, as well as form and function.

The term **evolution** here is used with caution, because it has a variety of meanings. It means different things to different people. Evolution can mean 1) change within a kind over time; 2) change from one kind to another kind due to the Darwinian mechanism of random variation and natural selection; and 3) descent with modification (all life forms being related).

Proponents of both naturalistic descent and creation models believe in the first definition of evolution that is biological change within kind (microevolution). Views two and three above involve macroevolution, a process that has never been observed in nature. These views differ in the amount of evolution, or biological change, that has taken place in microorganisms, plants, and animals over time.

Supporters of naturalistic descent see man as related not only to other primates but also to bacteria, protozoans, fungi, plants, and invertebrates. In the naturalistic view, the mechanisms of natural selection, mutations, immense periods of time, and chance physiochemical laws are assumed to be sufficient to explain the vertical, developmental macroevolution of all life forms.

In contrast, proponents of intelligent design while agreeing that current life forms may differ somewhat from their ancestors eons ago, believe the evidence from the fossil record and anatomy favor a model of microevolution (one with limited change or stasis) rather than macroevolution. Yes,

Table 2.1. Predictions about the nature of the human body according to contrasting views of origins.

Category	Predictions of Naturalistic Descent	Predictions of Intelligent Design
Anatomical Structures:	Low level of order Many are vestigial	Highly ordered No "useless organs"
and/or Body Parts:	Useless organs that are "leftovers" from evolution Many imperfections	All have a function Perfect in original design
Development:	Several vesitigial organs "Ontogeny repeats phylogeny"	Organs fully functional Purposeful ontogeny Versatile use of embryologic stages for design Nothing disposable
Genetics/ Inheritance:	Blending/gradual Transitions between kinds	Mosaic design Distinct traits Kinds separate
Mutations:	Occasionally beneficial	Basically harmful or lethal with only a few that are beneficial in certain environments
Natural Selection:	Creative process	Conservation process Device to eliminate harmful mutation
Nature of Man:	Slightly different from other primate "Misfits are disposable"	Distinctive from and superior to all primates There are no "misfits"; all have value
Organization:	Chaotic at times	Ordered most of time
A Product of:	Tinkering by nature	Divine Artificer (Craftsman)
Relationship among Body Parts:	Independent parts Mechanistic properties	Interdependent parts Coordinated properties

some species may have undergone speciation (within their kind), such as in bacteria, Drosophila (fruit flies) or Galapagos finches. The evidence for microevolution (limited change within a kind) is compatible with the creation view.

Intelligent Design

In 1802, William Paley, in his book *Natural Theology,* made the observation that if you were walking through the woods and you were to see two objects, a stone and pocket watch, lying on the ground and asked yourself about the origin of these objects, you would not hesitate to say that the watch was a product of a watchmaker and the stone was likely a product of weathering, erosion and fragmentation of a larger rock. As you studied and looked at the watch gears, springs, and screws, you most likely would deduce the watch was produced by conscious design and the handiwork of a watchmaker. You would not deduce this about the stone, because its surface and features are much more random.

Paley provided the first substantial logical argument for intelligent design theory. This is the concept that sees living organisms as the product of careful and conscious design and creative acts. A close examination of life forms reveals the details of structure and function so well and perfectly formed that they cannot be explained by chance physiochemical laws alone. The watch shows clear evidence of organization because of the way the components are put together to achieve a purpose.

The inference is inevitable that the watch must have had a Maker. Paley (1802) argued that like the watch's arrangement, the anatomy of the human eye, or the hand with its opposable thumb, logically demand a Divine Artificer (Craftsman) because of their complexity. The mechanisms proposed in macroevolution are insufficient to explain the seemingly perfect design in the human body.

How High Can You Jump?

Suppose your neighbor came to you saying that he had been exercising and wanted to impress you with his athletic ability. "Today, I cleared four feet in the high jump." You would have no reason to doubt him. If on successive days, he came to you bragging of his new athletic prowess being a foot higher each time, then you would probably start wondering about the truth after he claimed he cleared more than six feet. If he were one of the best athletes in the world (like Olympian Carl Lewis), then a seven foot leap would be as far as you could possibly trust him. If he qualified himself and said he used a pole vault to complete the jump then you might believe him up to a point, perhaps as high as twenty feet.

If your neighbor kept coming to you to brag about his jumping to a mountain top, such as jumping on top of Mt. Rushmore and standing on top of the carved President heads, you would not just harbor doubt, but figure him to be a liar or lunatic. If he wanted to convince you that he was not telling a fairy tale, he might qualify himself and tell you that he used scaffolding in several leaps. Then you would want to know where his scaffolding had gone after the jumps. Perhaps he would say, "Oh it disappeared!" If you pressed him further to provide evidence of this scaffolding, he tells you that he took it apart and used it to make his son's tree house. His story is

Figure 2.1. Mount Rushmore.

very convenient. Because it is difficult to prove him a liar, you change the topic of conversation to the NBA championship series or some other topic.

This allegory of high jumping can be applied to evolution. Microevolution refers to changes that can be made in one or a few simple jumps, whereas, macroevolution refers to changes that appear to require large jumps. Darwin's proposal that relatively tiny changes can occur in nature was an important conceptual advance.

There seems to be strong evidence that one or two species of Galapagos finches changed into clusters of species. We observe bacteria evolving from the antibiotic sensitive condition into new strains that are resistant to antibiotics. Rhinoviruses, those that cause the common cold may owe their diversity to mutation, selection, and diverse immune systems of the human body (Gillen and Mayor, 1995). Viral antigens do change over time. Dawkins' proposal, however, in *The Blind Watchmaker* (1996a) and *Climbing Mount Improbable* (1996b), insists that entire new organs, organ systems, and species have come about by means of numerous mutations acted upon by natural selection over long periods of time. This is like your neighbor who claimed to have climbed Mt. Rushmore with a scaffolding that no longer exists.

Assumptions Color Your Conclusions

As you think about the origin of man and his body, you should be aware that your assumptions color your logic and eventually your conclusions. In math, your assumptions determine your conclusions. If someone asserts that $4 + 4 = 13$, you would probably think he failed arithmetic and was extremely naive. Imagine instead that he qualified his explanation and proceeded to tell you that he was working in Base 5 and then explained to you what he meant by 13 was $(1 \times 5) + (3 \times 1)$. Such a person actually knows more about math than the average individual.

In biology today, most scientists insist that you must work within a naturalistic, evolutionary framework. To "politically correct" scientists in the 1990s, creation science and intelligent design are not an option. If you do not work within the neo-Darwinian framework, then you are considered ignorant, naive, or just uninformed. This bias is like the mathematician who insists that others can work only in Base 10 and cannot think in terms of Base 5 or any other system that is not considered acceptable. If you assume a Base 10 mathematical logic, then your answer to $4 + 4$ *must be* 8.

If you allow for other base systems in math, then you are open to other "truths." Likewise, if you keep open to the design option of the human body, you will actually see that the data support it. You might find it surprising, but the truths found in the Bible are corroborated by good science and serious investigation into anatomy and physiology. Remember one principle from this section, that assumptions color your thinking!

Summary

In summary, the Darwinian naturalistic descent view and Paley's intelligent design theory provide two very different perspectives on the development of the human body. Table 2.1 summarizes these two views of origins. You will want to compare the predictions of these two views and evaluate the data that support each one.

According to Darwin, nature tinkered for a long time and human body parts were the result. For Paley, a Divine Artificer (Craftsman) was responsible for the design seen in human anatomy. Which one is true? You are challenged to make a choice:

Choose you this day whom ye will serve; whether the gods which your fathers served that were on the other side of the flood, or the gods of the Amorites, in whose land ye dwell; but as for me and my house, we will serve the Lord.

Joshua 24: 15

Today's choice can be compared to a decision the ancient Israelites had to make between Jehovah or the Amorite gods. Today a worker in science must choose between the Creator and the blind forces of macroevolution. I believe that after all the evidence is examined, you will find new confidence in the biblical summary of creation:

Through faith we understand that the worlds were framed by the word of God, so that things which are seen were not made of things which do appear.

Hebrews 11:3

Before one can have the Hebrews 11 faith, one must understand God as the Creator. Creation came after God (Jesus) acted. This is the foundation of that faith.

Chapter Three
Human Cells and Development

The secret of (human) membership lies locked away inside each cell nucleus, chemically coiled in a strand of DNA.

Brand and Yancey (1980, p. 45)

The most basic unit of the body is the cell. One of the fundamental discoveries of the 20th Century is deoxyribonucleic acid (DNA). DNA is found in the nucleus of every human cell. It serves as a "blueprint for life." Observations of living cells confirm that most of the development in the human body is both coded for and controlled by DNA.

The mother secretes another nucleic acid called ribonucleic acid (RNA) into the egg. RNA from the mother directs the stages of development in the embryo. DNA consists of a simple but elegant pattern. The DNA bases adenine (A), thiamine (T), cytosine (C), and guanine (G) are the parts of DNA that code for cell structure. Like a blueprint for a building, these same bases determine the details of the body form and function.

Four themes that can be seen in development in human cells, organs, and organ systems include 1) many living organisms and their parts can best be described as mosaics, thus, comparable to artwork; 2) each body cell clearly illustrates a coordinated complexity; 3) many cell molecules and body organs illustrate an efficient adaptation to their environment; and 4) each body part is fully functional.

Fluid Mosaic Model of the Cell

Many living organisms and their parts can best be visualized as mosaic patterns. Mosaics in nature mirror artwork. The tessellations of Escher represent a mosaic, blending mathematical principles with an artistic vision. Both have patterns and designs in which contrasts exist between colors, pigments, and structures.

Biological mosaic patterns are most readily seen in cell structure. The cell membrane is a collage of many different proteins embedded in the fluid matrix of the lipid bilayer. Proteins in and on this bilayer have been described by some as "sailboats on a lake." The lipid bilayer is the main fabric of the cell membrane, but the protein composition (within the bilayer) determines the specific functions of most cell membranes. The cell membrane and other intracellular membranes have their unique sets of proteins.

For example, more than 50 protein kinds have been found in the plasma membrane of red blood cells. In fact, biologists now refer to the best model of the cell membrane as the **Fluid Mosaic Model**. This model (Figure 3.1) describes the cellular proteins embedded in mosaic patterns within a lipid bilayer. This newer model reflects a change from the original Davson-

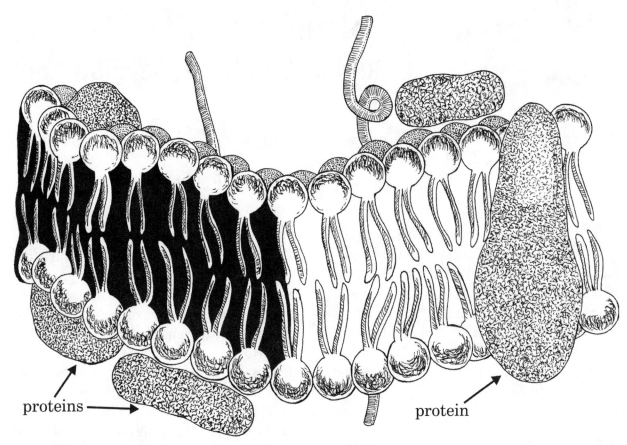

proteins

protein

Figure 3.1. Fluid Mosaic Model of the cell membrane.

Danielli "sandwich model" where the lipid layer was squeezed between two protein layers. All this intricacy fits with the creationist concept that the cell is the result of a Divine Craftsman skillfully planning beautiful mosaics in membranes in order to transport and to selectively screen substances across this important boundary.

Irreducible Complexity of the Cell

Each body cell clearly illustrates a coordinated complexity. This evidence has been compiled by prominent scientists Drs. Behe and Denton (Denton, 1986).

Biochemist Michael Behe, author of *Darwin's Black Box*, points out that *"the*

simplest self-replicating cell has the capacity to produce thousands of different proteins and other molecules, at different times and under variable conditions" (p. 46). Synthesis, repair, communication-all of these functions take place in virtually every cell.

According to Behe (1996), many processes in the human body exhibit **irreducible complexity**. An irreducibly complex system is one that requires several interacting parts to be present and functioning at the same time, where the removal of one (or more) of the parts causes the whole system to malfunction. Destroy one part and the whole falls apart—losing any beneficial adaptation or selective advantage. The purported mechanism of evolution, on the other hand, is that a new trait will confer a **survival advantage,** and thus enable its possessors to compete better than organisms without the trait. In macroevolution, any new trait would have to be completely developed; no half-way measures would do. Given this requirement, new features are so complex that neo-Darwinian gradualism is very improbable because an incompletely developed trait would confer no selective advantage.

Many cellular processes will not work unless every part is present and functioning. One such process is the transport of proteins across the cell membrane. Proteins do not just float around freely inside of cells. To paraphrase Michael Behe (1996, p. 101), eukaryotic cells have a number of different compartments, like rooms in a house. When a protein is made it has to get from the compartment where it's made to the compartment where it's supposed to be.

Consider what is involved in simply moving a protein through a compartment wall. Cells have two fundamentally different ways of doing this: gated transport and vesicular transport. In gated transport, the compartment wall is equipped with a "gate" and a chemical "sensor." If a protein bearing the right "identification tag" approaches, the sensor opens the gate and allows the protein to pass through. If one with the wrong tag comes near, the gate stays shut. These processes can be clearly observed in the largest internal organ of the body, the liver. One of the main functions of the liver is to control the level of vital nutrients in the blood, such as carbohydrates and proteins (Table 3.1).

The importance of the liver should not be underestimated in the body. Liver transplant patients know their lives hang in the balance until a fully functioning liver replaces a degenerate one. Liver cells provide an example of vesicular transport of proteins and carbohydrates in an organelle, the smooth endoplasmic reticulum (ER). On their ribosomes, liver cells make many blood proteins that are transported in vesicles through the ER before release into the bloodstream.

In addition, liver cells store and transport carbohydrates in the form of glycogen. The hydrolysis (splitting) of glycogen leads to the transport and release of glucose subunits from the liver cells, a process that is important in the regulation of blood glu-

Table 3.1. Functions of the Liver (Campbell, 1996).

1. It detoxifies otherwise poisonous substances.

2. It produces bile, as well as enzymes in metabolism.

3. It removes glucose from the blood under the influence of insulin and stores it as glycogen. When the glucose level falls, the hormone, glucagon, causes the liver to break down glycogen, releasing glucose into the blood.

4. Liver cells make many complex blood proteins.

5. Liver cells convert nitrogenous wastes into urea that can be excreted by the kidneys. The liver, along with kidney, helps to regulate the contents of the blood.

cose levels. Movement of glucose within and between cells is facilitated by vesicular transport. In addition, enzymes of the smooth ER of liver cells help detoxify drugs and poisons. Detoxification usually involves adding hydroxyl groups (-OH) to drugs, increasing their solubility, and making it easier to flush the compounds from the body. The removal of alcohol, barbiturates, and hazardous drugs in this manner by smooth ER liver cells is also accomplished through vesicular transport.

Note all three components-the gate, the sensor, and the tag-must be in place for transport to occur. Most biochemical systems in the cell are much more complex than this. The total cooperation of many organelles produces an effect greater than the sum of the individual organelles. Biologists refer to this cooperative effect as an **emergent property** of the cell. Do you think this cooperation happened by chance?

Marks of Design

Michael Denton, a molecular biologist (who has both M. D. and Ph. D. degrees), wrote a blockbuster book in the 1980s called *Evolution: A Theory in Crisis*. He says that the "puzzle of perfection" in living things critically challenges any theory of evolution that is based on chance mutations. New discoveries in biology are opening up new levels of detail in the workings of the human body and are revealing ever more precise functioning.

Every cell type and organ in the body is necessary and useful, and within each organ there is an incredible inner precision. The liver, the body's largest internal organ, is a multifaceted, vital organ - one

without which people cannot live long. Indeed it is only one of many organs, which when working properly, illustrate great efficiency and order in complexity.

Another "cell team" that illustrates the "emergent properties" concept and a "puzzle of perfection" (Denton, 1986) is the immune system. The actions of the immune system come from the cooperation of many cells, such as monocytes, lymphocytes and macrophages ("big eaters" of harmful chemicals and pathogens). The ability of macrophages to recognize, apprehend, and destroy bacteria depends on the coordinated activity of many cells. The cytoskeleton, lysozymes, and the plasma membrane function in phagocytosis. Like many other bodily functions, the immune system depends on the interactions of many cells and cell parts.

Dr. Denton also discusses new discoveries in human physiology that reveal even more precision at the molecular level in various organs. Every organ in the body is necessary and useful, and within each organ there is an incredible inner order that provides precise coordinated functioning. Today's scientific tools of investigation are much more powerful than any during Darwin's day. Scientists can now explore most of the complex, interlocking biochemical systems.

As cell biologists discover numerous cascades of biochemical reactions, they have learned that these reactions act in perfect harmony. The human body is able to facilitate these activities when they are needed and it can inhibit them when they are not needed. The sequence of commands is programmed in the DNA code that directs almost all cellular processes. Such microscopic precision is dazzling.

It is a mathematical absurdity to suggest that billions of complex cells could all coordinate their precise molecular activities by blind chance. The inadequacy of the theory of body organs arising by chance is well illustrated in the human kidney. There is an elegance of design manifest in the kidney. It combines many wonderfully clever adaptations to control water and electrolyte homeostasis, as well as blood pressure. At this same time, the kidney concentrates and eliminates wastes from the body in urine.

In Chapters 4 to 8, we present several other examples of intricate design, including the structure of the hand, the impressive immune system, the extraordinary excretory system, the dynamic digestive system, the blood clotting mechanism, and the biology of vision. Neo-Darwinism falls short, lacking a scheme of origins. Because there is no simpler level to which they can appeal, macroevolutionists cannot explain how these systems arose. According to Behe (1996), there are no detailed mechanisms for the origin of living cells.

Drs. W. Bradley and C. Thaxton (1994) carry this criticism of macroevolutionism over the study of biological code chemicals. Unlike the complexity of a snowflake that derives its structure from the nature of the materials it is made from, the complexity of genetic information is independent of the nature of nucleic acids that record it. The nucleic acids are merely the vehicle by which this marvelous biological information is carried. Intelligent causation once again is strongly suggested.

Human Development: Every Part Essential

There are certain areas of human genetics in which researchers expect to expand their current knowledge. Two of these are the regulation of gene expression and the mapping of the entire human genome. The role of the vast majority of DNA that has not yet been assigned a function has been inappropriately named "junk" DNA. Research published in early 1992 demonstrated that an intron, which is thought to be a non-coding region of DNA, plays a role in the function of transfer RNA, which is critical to protein synthesis. Additional research may reveal other functions of this so-called "junk" DNA (Access Excellence, 1996). Indeed, each genetic component seems to be essential for body function. Even the so-called "vestiges" are fully functional organs. In human development, there are few (if any) wasted parts in the fetus and embryo. The concept of there being useless or vestigial body parts is an outdated idea that has largely been disproved (Bergman and Howe, 1990).

Recapitulation Revisited

Ernst Haeckel, a German anatomist first proposed the recapitulation idea that ontogeny recapitulates phylogeny (development replays macroevolution) over 100 years ago. Better known as the "biogenetic law," it claims that in its development each embryo passes through abbreviated evolutionary phases that resemble the developmental stages of its ancestors. Some suggest that man evolved from fish and amphibians, which have gills, and then point to the "gill slits" which human embryos are said to have in one stage of their development. Evolutionists base the recapitulation idea on the faulty assumption that any part superficially resembling another in animals must have a common origin.

This idea was dismissed for a number of years by professional biologists but it keeps appearing. Dr. Ken Miller (1996), for example, has recently tried to give credibility to this nineteenth-century thinking in his recent *Life* magazine article. This article, in a speculative essay format, contains beautiful photographs of vertebrate embryos, intended to convince readers of its scientific merit.

Does ontogeny repeat phylogeny? No, not really, says Dr. Jonathan Wells, a noted Ph.D. embryologist (Wells and Nelson, 1997). Wells, at a science meeting, says Miller's evidence for the recapitulation idea is the "Piltdown" hoax of 1996 embryology. The following evidence stands against Haeckel's biogenetic law:

- The so-called "yolk sac" is really a blood sac. Blood cells originate in the structure. Later, bone marrow develops from this tissue in the human fetus.
- The so-called "gill slits" are really wrinkles in the throat region. This body tissue becomes the palatine tonsils, middle ear canal, parathyroid gland, and thymus in humans.
- The so-called "tail" is an attachment site for coccyx muscles. It is necessary for good posture and support, and aids in defecation.

In fact, neither gills nor their slits are found at any stage in the embryological development of any mammal including man. These folds in the neck region of the mammalian embryo are not gills in any sense of the word and never have anything to do with breathing. They are merely inward folds, or wrinkles, in the neck region resulting from the sharply down-turned head and protruding heart of the developing embryo. The folds eventually develop into a portion of the face, inner ear, tonsils, parathyroid and thymus. None of the reputable medical embryology texts that we checked claimed that there are "gill slits" in mammal development.

To say that people have a "tail bone" is to assume human evolution from a tailed, vertebrate ancestor. Anatomically speaking, this bone is the coccyx. It has nothing to do with the tail that we usually associate with monkeys or dogs. In the developing human embryo, before it is covered with important muscles of the lower intestine, the coccyx only looks like a tail. There are a number of embryonic structures that superficially look like something else. Evolutionists use these developing structures to make their case for macroevolution. However, science has shown each structure to be important for continued human development. In this case, the coccyx is the base of the spinal column. It is the site of attachment for certain lower intestinal muscles needed for upright human posture, and it aids in defecation.

Still, the "gill slit" myth is perpetuated in many college biology textbooks as evidence for macroevolution. The fictitious "gill slits" of human embryos discussed by Haeckel, for example, are supposed to represent the "fish" or "amphibian" stage of man's evolutionary ancestors. Much of Haeckel's work has been shown to be a fraud (Pennisi, 1997).

Most professional embryologists no longer believe this "gill slits" myth of the biogenetic "law." Even evolutionist Dr. P. Ehrlich said:

...this interpretation of embryological sequences will not stand under close examination. Its shortcomings have been almost universally pointed out by modern authors, but the idea still has a prominent place in biological mythology.

Menton, 1991, p. 2

Embryologist Dr. E. Blechschmidt reveals some of his frustration with the persistence of this myth:

The so-called basic law of biogenetic is wrong. No buts or ifs can mitigate this fact. It is not even slightly correct or correct in a different form. It is totally wrong.

Menton, 1991, p. 2

Yet, in most introductory college biology textbooks, embryological recapitulation is offered as evidence for evolution. Since the time of Darwin, the argument from similarity comprises much of the case for the general theory of evolution. It is assumed that similarity provides evidence for an evolutionary relationship and the degree of similarity predicts its proximity in time. Neither of these propositions is supported by scientific data.

Chapter Four
Structure and Function: A Planned Relationship

Thus, by ocular observations and dissections, by experiments and measurements, and by calculations and inductive reasoning, it is absolutely necessary to conclude that the blood is propelled by the heart in a circle and this is the only end of the heart.

William Harvey, M.D., 1628 (Eakin, 1982, p. 9)

Given a choice of dissection tools, you would not make an incision in a preserved specimen with a blunt probe nor would you use a scissors to separate delicate blood vessels that are attached. Ideally, you would use a sharp scalpel to make an incision and a blunt probe to separate the blood vessels in your dissection. The structure of an instrument determines its function.

In like manner, structure and function are correlated at all levels of biological organization. This theme is a guide to anatomy at its many structural levels, from molecules to organisms. Analyzing a biological structure gives us clues about what it does and how it works. Conversely, knowing the function of a structure provides insight about its construction.

The Remarkable Fit of Body Tissues

The principle that structure is the basis of function applies to the cells and tissues as well. **Histology** is the study of tissues, which are defined as a group of similar cells forming a definite and continuous fabric, usually having a comparable function (Marieb, 1994). The function of a particular tissue is limited by its structure, as illustrated by epithelium. It can do only as much as its structure will "allow." The structure of each type of epithelium fits its function. Some of the major types of epithelium from four regions of the body are depicted in Figure 4.1.

Pseudostratified columnar epithelium (I) has cells (like bricks on end) with relatively large cytoplasmic volumes. It is often located where secretion or active absorption of substances are important functions, like those found in the trachea. Dirt or foreign particles are escalated out of the body through cilia and mucous.

Simple cuboidal epithelium (II) is specialized for secretion and makes up the epithelia of the thyroid gland shown. These cells (like sugar cubes) secrete hormones into the bloodstream. In contrast, **simple squamous epithelium (III)** is a relatively leaky tissue. These cells (like floor tiles) are specialized for the exchange of materials by diffusion. These epithelia are found in such places as the linings of blood vessels and air sacs (alveoli) of the lungs.

A **simple columnar epithelium (IV)** has cells with relatively large cytoplasmic volumes, and is often located where secretion or active absorption of substances are im-

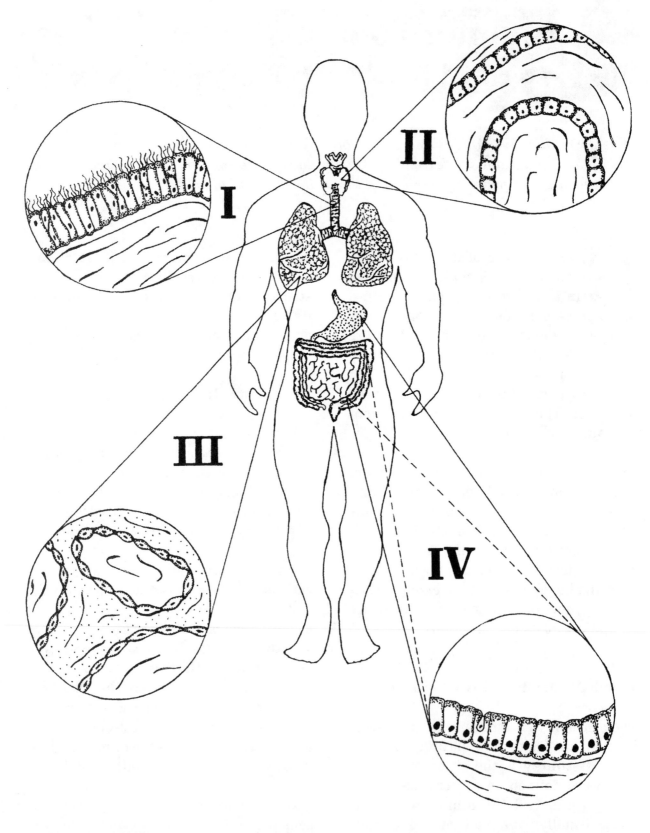

Figure 4.1. Epithelial tissue types in the human body: pseudostratified columnar epithelium (I), simple cuboidal epithelium (II), simple squamous epithelium (III), and simple columnar epithelium (IV).

portant functions. The stomach and intestines are lined by a columnar epithelium that secretes digestive juices and absorbs nutrients.

Bones: A Study in Design Engineering

One of the more familiar types of connective tissue to the anatomist and layperson alike is bone tissue. Perhaps, the simplest anatomy to understand is the skeleton, and the easiest places to observe the relationship of form to function are its individual bones. Anatomy and physiology students are impressed with the robust, intricate, sculpted design of bone. Why is this? Perhaps it is due to the function and structure of bone, the interaction and fit of various bones much like mechanical objects. Fossilized bones, too, have an ancient mystery surrounding them, their very existence is explained by one's world view. Convinced that "form follows function," we will first consider the functional importance of bones with regard to our physical makeup and survival.

Bones will be examined not only for their mechanical function but also the role that the largest and smallest bones have in our anatomy and physiology. Since human skulls (mostly shattered and incomplete) and teeth are some of the most important fossil specimens used to speculate regarding the strange fiction of "human evolution," we will look at these too. Special attention will be given to the dynamic nature of bone as a living, growing tissue. We will see how our endoskeleton grows in response to the physical demands an individual places upon it.

An ultrastructural study of bone will show

special cells that either break down, maintain, or add bone deposits in a very detailed and orchestrated fashion according to daily physical demands. These cells must work together in a precisely coordinated fashion. An imbalance of the activity of these cell types will result in serious conditions like osteoporosis.

The Human Foot: Marvel of Engineering

The human foot is a marvel of engineering. The foot is highlighted by its specially designed arches and series of bones that allow one to either pound the gridiron or balance tip-toed during a long "Swan Lake" performance. You will be amazed at the design features of this complex, highly integrated system of dynamic tissue called bone.

Each foot (pes) is composed of an ankle, instep, and five toes. The tarsus (ankle) consists of seven tarsal bones that form the heel and the posterior portion of the instep. The instep is composed of five elongated metatarsals that form the ball of the foot and articulate with the tarsus. Ligaments connect the tarsals and metatarsals to form the arches of the foot.

A longitudinal arch stretches from the toe to the heel. The transverse arch extends across the foot. The arches provide a stable, springy support for the body. Design is seen even in the small details. Each toe has three phalanges, corresponding to those phalanges of the fingers, except for the big toe, which has only two phalanges and provides precise movement and balance in the foot.

The foot has two arches that support the

weight of the body and provide leverage when walking. These arches are formed by the structure and arrangement of the bones held in place by ligaments and tendons. The arches are not rigid; they yield when weight is placed on the foot and spring back as the weight is lifted. The longitudinal arch is divided into medial and lateral parts. It is supported by the calcaneus proximally and by the heads of the first three metatarsal bones distally. The wedge, or keystone, of this part of the longitudinal arch is the talus. The shallower lateral part consists of the calcaneus, cuboid, and the fourth and fifth metatarsal bones.

The transverse arch extends across the width of the foot and is formed by the distal part of the calcaneus, navicular, and cuboid bones, as well as the proximal portions of the five metatarsal bones. The weakening of the ligaments and tendons of the foot decreases the height of the longitudinal arch in a condition called the pes planus, or flat foot (Van de Graff, 1998).

The foot is the humblest member of man's anatomy. The human foot with its five toes is a miracle of construction. It consists of 26 separate bones of various sizes and shapes bound together by a system of ligaments. It is supported by a complex array of muscles and supplied with a network of fibers and blood vessels. The different bones articulate in gliding joints, giving the foot a degree of elasticity and a limited amount of motion. The arrangement of the bones is such that it forms several arches, the most important of these being the long arch from the heel to the ball of the foot.

These arches are held in place and sup-

ported by a complex of strong muscles to carry the weight of the body just as the steel cables carry the load of a suspension bridge. The construction also gives elasticity to the foot, making walking, running, and other movements possible. The ability to walk and to move from place to place and to balance the tall upright body on a comparatively small pedestal is itself a most remarkable feature of the human body. If the foot were flat and rigid, fixed at right angles to the bone of the leg, walking would be difficult or impossible. The elastic arches also serve as shock absorbers to soften the jar resulting from walking on a hard surface.

The human foot is a masterpiece of engineering. It is a miniature suspension bridge but more complicated than an ordinary bridge. Would anyone say that the Golden Gate suspension bridge just happened? Of course not, if he were truthful! But why do people assume that an even more intricate mechanism of the human foot could have just happened without intelligent cause or the workmanship of a Master Engineer?

To add to the wonders of the human foot, we must also remember that it has been reproduced billions and billions of times in every human birth with exactly the

Figure 4.2. Golden Gate Bridge, San Francisco, California.

same shape and form and with the same number of bones and tendons and nerves. It has both strength and agility, just like the famous Golden Gate Bridge, to handle varying weather conditions. And so we see that even the humble foot of man is one of the wonders of his body that glorifies the wisdom of the Creator who made it (Parker, Graham, Shimmin and Thompson, 1997).

Form and Function in Skeletal Muscles

Another example of this structure and function theme is the biceps brachii muscle of the upper arm and its control by neurons. Long extensions of the neurons transmit electrochemical impulses, making these cells especially well structured for communication. The movement of the body depends on the structure of muscles and movements of bones. The movement produced by a contracting muscle depends on how it is attached to the bones and how the bones articulate with each other. In such a familiar example, the relation between structure and function is obvious.

The dependence of activity on structure becomes more subtle, but no less real, as we direct our attention to the lower levels of organization, such as tissues, cells, and organelles. Our understanding of how a muscle contracts rests on the ultrastructure of the contractile machinery as well as on its chemical properties.

As an example of functional anatomy at the subcellular level, consider the organelles called mitochondria. Mitochondria are quite numerous in skeletal muscles. They are the sites of oxidative metabolism, which powers the cell by a process using oxygen and tapping the energy stored in glucose and triglycerides. A mitochondrion is surrounded by an outer membrane, but it also has an inner membrane with many infoldings. Molecules embedded in the folds of the inner membrane accomplish many of the steps in oxidative metabolism. The infoldings pack a large amount of membrane into a minute space. In exploring life on its different structural levels, we discover operational beauty at every turn.

Structure and Function of Vital Organs

Two organs that man depends upon most for daily living are the heart and the lung. Both of these vital organs clearly show evidence of a planned relationship between their structure and function. In the early days of science, perception of how these organs functioned was inadequate in terms of the knowledge we have today. Before the days of Harvey and other Renaissance scientists, the form and function of these vital organs was merely hypothesized. Experimentation later corrected early misconceptions.

Harvey and the Heart

The heart of creatures is the foundation of life, the Prince of all, the Sun of their microcosm, on which all vegetation does depend, from whence all vigor and strength flow.

William Harvey, MD, 1628
Letter to Charles I of England
(Nuland, 1997, p. 208)

It was not until the early 1600s that William Harvey gave the explanation of blood circulation that we recognize today. According to Galen (131–210 AD), the blood was formed in the intestines, and then in the liver was charged with *natural spirits* and finally carried to the right side of the heart by the veins. Contractions of the right atrium and ventricle were thought to cause the blood with natural spirits to ebb and flow throughout the veins to all parts of the body. Some of the blood was supposed to ooze by invisible pores into the septum of the heart and into the left ventricle where it mixed with a certain quantity of air. Last, the blood was thought to be drawn from the lungs though the pulmonary veins and stuffed with vital spirits. (Galen thought that all living things were controlled by a mystical influence that he called "vital force.") Pulsations of the left ventricle then were thought to move the vital spirits back and forth in the arteries to heat, quicken and restore all body tissues (Morrison, Cornett, Tether, and Gratz, 1977).

In contrast to Galen, William Harvey's view of the heart and circulation rested neither on books nor opinions of his predecessors, but on direct ocular observa-

Figure 4.3. William Harvey (1578–1657).

tions, dissections, and experimentation. In 1628 Harvey published a small book written in Latin called *Exercito Anatomica de Motu Cordis et Sanguninis in Animalibus*, which is translated as *An Anatomical Dissertation on the Movement of the Heart and Blood in Animals*.

Harvey's idea had been by far the most widely accepted of several competing theories of blood flow, but *De Motu Cordis* (as it is usually called) would in time bring the death blow for all of them. Harvey used a combination of quantitative and qualitative methods to make a series of easily confirmable observations and measurements that led him to his conclusions. He proclaimed in one striking sentence the entire contents of his Chapter 14 in this monumental book:

It must be of necessity concluded that the blood is driven in a round by circular motion in creatures and that it moves perpetually; and hence does arise the action and function of the heart, which by pulsation it performs; and lastly that motion and pulsation of the heart is the only cause.
                                    (Harvey, 1995, p. 91)

Note that Harvey summarized the reason for the heart's design was to accomplish the purposeful function "*and lastly that motion and pulsation of the heart is the only cause.*" With this, he no doubt meant to lay to rest the misguided confidence of his readers in those mystical attributes of the heart so dearly cherished by church people, kings, and commoners alike. Perhaps his message was not lost on Charles, though the royal patient must have taken consolation in knowing that his doctor still believed so strongly in cardiac supremacy that he compared the heart's indispensability to that of the king himself. (It should be noted that Charles was the head of the Anglican church, in theory, and described historically as a defender of the faith. Harvey also recognized this same faith of the king and his writings imply that he also was both an Anglican and a creationist.)

Although the implications of Harvey's contribution were not at first universally appreciated (and in some quarters not even accepted), the publication of *De Motu Cordis* opened the way to a profusion of studies aimed at elucidating aspects of cardiac function and the details of the various elements, such as blood pressure, the pulse, and the activity of cardiac muscle. Finally, in the late eighteenth century, physicians began to use the fruits of findings based on Harvey's principles to help them understand the diseased hearts of their patients. By the first decade of the nineteenth century, the technique of percussion had been developed, by which it was possible to estimate the size of the heart by tapping on the chest. In 1817, the stethoscope was invented and cardiology advanced further (Nuland, 1997).

Harvey had built upon the anatomical approach of Theophrastus Hohenheim, who wrote under the name of Paracelsus (1490–1541). Paracelsus believed that if the structure of the body was to be learned, direct observation was essential rather than textbook anatomy (Morrison *et al.*, 1977). The combined work of these two men revolutionized the study of anatomy, thereby advancing the way science and medicine were done (McMullen, 1998).

The importance of Paracelsus' work was that it clearly demonstrated that the flow of blood through the body is really a circulatory movement. This movement flows one way through the arteries, capillaries, and veins, all forming conducting channels. Translated from Latin, Harvey said:

The blood is propelled by a muscular pump, the heart. Blood flows into a chamber of the heart. Then, the entrance is closed by a valve and the chamber is contracted, forcing the blood into a second chamber and then into an artery.

(Eakin, 1982, p. 9)

Like Isaac Newton, Harvey believed the study of the human body would reveal the "fingerprints of God." Harvey sometimes referred to the intricate and orderly circulatory patterns in the heart as the "fabric of life." Other scientific "men of God" during this era included Leonardo da Vinci, Blaise Pascal, Anton van Leeuwenhoek, Robert Hooke, and John Ray. These were men who believed the inspiration and authority of the Bible (Morris, 1988). Newton is frequently quoted, in regard to his study of science, as *"thinking God's thoughts after Him"* (Morris and Morris, 1996). Their studies were an extension of their faith in God.

During this time period, it is remarkable that Harvey used the scientific method of exact experimentation and mathematical calculation to support his theory. He had no microscope capable of revealing the capillaries, but he reasoned that there must be such a connection between arteries and veins. Like the Darwinists of today, many people in Harvey's day refused to accept his explanation, especially those who favored the established ideas of Galen. Capillaries were later demonstrated by means of a light microscope in 1632 by Marcello

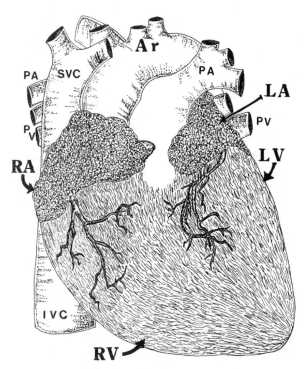

Figure 4.4. The human heart: right atrium (RA)—the "ear" of the heart, right ventricle (RV), left atrium (LA), left ventricle (LV), aorta (Ar), pulmonary vein (PV), pulmonary artery (PA), superior vena cava (SVC), and inferior vena cava (IVC).

Malpighi and this vindicated Harvey (Morrison *et al.*, 1977).

As Harvey discovered, the heart is actually two pumps, right and left, with a septum between the two sides so that the blood does not go from one side to the other. The left pump is somewhat larger and stronger than the right pump. Its function is to receive blood from the lungs in its upper chamber, or atrium, and send it out all over the body from its lower chamber, the ventricle. If you look at the heart of a preserved specimen you will note that the atrium has very thin walls. When it is contracted, it is very small. Notice the correlation of structure and function; the ventricles are designed to pump blood over far distances to the lung and entire body, as indicated by their more "muscular" cham-

bers. Atria are designed for receiving venous blood and pumping it only downwards only a short distance, to the larger, stronger ventricles. Hence, atria have thinner walls.

Harvey found the vessels involved in blood circulation also show a planned form and function relationship. When examining a cross section of a vein, capillary, and artery, one can see a single layer of tissue called the endothelium. In veins and arteries, the endothelium is surrounded by a muscular layer and a protective layer of connective tissue. The arteries have a thick muscle layer to aid in pushing blood out to the body tissues farthest from the heart. The layers of the veins are thinner and more flexible than those of similarly sized arteries and this fits their function of returning blood to the heart.

Another important difference between veins and arteries is that veins in the lower body contain valves that are flaps of tissue that help keep blood flowing in only one direction by preventing back flow. These valves open under pressure of blood going toward the heart, and close when the blood begins to go backward in response to the pull of gravity. The walls of capillaries are the thinnest, having no muscular or connective tissue layer, and are designed for easy exchange of oxygen and carbon dioxide between the circulating blood and body tissues.

Thought Pattern of Harvey in Structure to Function

If you read the first English translation of *De Motu Cordis*, (first written in Latin, 1628) in describing much of his work in 1616, it reads a lot like the King James Bible not only in its stately Old English

Language but also in its logic. On more than one occasion, Harvey makes reference to God, the Divine, and refers to himself as a Christian in his writings. He had served as physician to King James I (authorized the KJV) and his son Charles I. With the coming of the Reformation and the availability of the Bible (1611) to all people, the commands of Christ had become apparent to all. It is no accident that a new interest and reform had come to science and medicine. Among Christians there was new interest and zeal in helping to heal the sick as Christ had commanded. William Harvey felt it his duty to his earthly and heavenly King to perform the best physiology and medicine as a Christian physician.

"The circulation of the blood is prov'd by consequence" (Chapter XVI) was his theme in developing the concept that one way valves control blood movement through veins from the heart to the outer arm. He used the logic that *"Consequence demanded a Cause."* Purposeful function demanded a designed structure, just as a first Cause was the logical antecedent to the human body and its circulatory system. Just as the effects of wind blowing on a tree infers a force in the wind itself. Harvey saw there must be a Supervisor to the orderly unidirectional flow of blood. It may have reminded him of the Levitical law (Lev. 17: 11) that describes the importance of blood giving life to all creatures and where its circulation is inferred. When Harvey conducted experiments on animals and himself, he observed blood flowing in one direction and noticed in some vessels a gate seemed to control the emission of blood. Later, he deducted logically that valves were the gated control that determined direction and quantity of blood flow.

Its structure determined its function.

It should be no surprise that Harvey made breakthroughs in physiology. Just as people were challenged to investigate the Bible for themselves during the reformation, and urged not to rely on Papal or church authority alone for Truth; men of science began to question Galen (and other ancient scientists) on what they wrote in ancient textbooks. They could investigate for themselves and see what the data (or observations) revealed. Speculation gave way to experimentation. Thus, Harvey broke with tradition and discovered not only that blood circulated throughout the

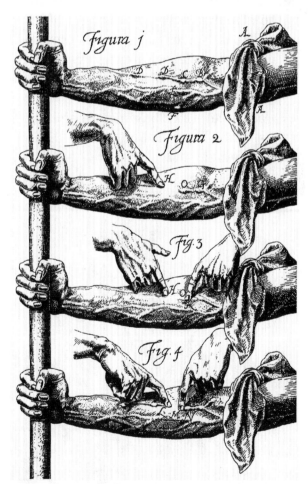

Figure 4.5. Harvey's design deduction on one-way valves in veins of the arm: An illustration from William Harvey's *De Motu Cordis* (1628).

body, but also that valves controlled the one-way flow of blood in the arms. If you read *De Motu*, the motivation is to both please the King (James I and later Charles I), as well to bring glory of God.

The Lung

The other vital organ that we depend upon each moment of life is the lung. People can hold their breath for a maximum of five or six minutes, denying the body gas exchange from the lung. If it were not for an automatic feedback mechanism in which the body starts breathing again during our unconscious state, the action of holding one's breath would ultimately lead to death. The basic unit of the human lung is the alveolus, which is a cluster of lung cells. The design of the alveolus provides:

- increased surface area for gas contact,
- moisture that aids in exchange and transport of oxygen,
- a one-cell-thick basement membrane,
- a site of oxygen and carbon dioxide exchange, and

- close proximity of the blood capillary bed to the air surface, promoting the absorption of gases.

In addition to structural adaptation of the alveolus in the lung, the diaphragm and intercostal muscles (below ribs) rhythmically contract to promote bulk inflow and outflow of air. Hairlike organelles called cilia in the respiratory tract remove mucous and harmful residues from the system. Once again, structure and function are highly correlated, demonstrating the optimal body plan for survival and adaptation in a changing world (Van de Graff, 1998).

Overlooked Anatomy

It is not unusual for diagrams in beginning anatomy and physiology texts to overlook some not-so-glamorous structures in the body. This is often the case because artists tend to copy diagrams from other books that are copies themselves. Instead of making figures from cadavers, in the case of humans, or specimens, in the case of animals, artists through the ages copied or modified "accepted" figures of anatomy. Frequently, one error reproduces itself for many generations of diagrams. Historically, this was the case for over a thousand years, until Vesalius and others during the Renaissance began to do cadaver dissection themselves. Today, the same types of errors take place. Therefore, it is always important to remember that nothing replaces first hand observation and thinking. Do not merely believe it because the text says so! For example, few textbooks depict the right auricle and the lingula of the left lung. These structures are sometimes affectionately referred to as the "ear of the heart" and the "tongue of the lung" by medical students. If we make

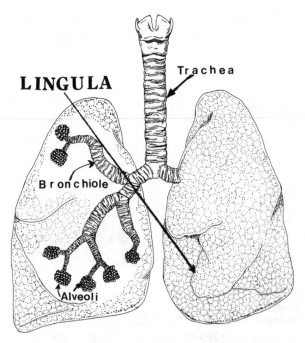

Figure 4.6. The human lungs, showing the lingula—"tongue" of the left lung.

an in-depth study of these structures that are frequently left out of texts, we will once again observe that structure and function are correlated. The *appendix aurculae* is so named due to its fancied resemblance to a dog's ear (Gray 1995, p. 435). It is a small conical, muscular pouch, the margins of which present a dentate edge. It projects from the sinus forward and to the left side, overlapping with the aorta. Next time you dissect a mammalian heart, like a beef heart, see if you can find the right atrium. This structure might also be called the **"ear" of the heart**. Another overlooked anatomical structure is the anteroinferior part of the superior lobe in the left lung. It forms a small tongue-like projection called the **lingula** (Latin for *tongue*) (Moore, 1980, p. 40). This structure, during cadaver dissections by medical students is frequently referred to as the **"tongue" of the lung.** We might note that it is the left lung that has a cardiac notch.

Both of these "appendages" afford protection over key blood vessels. They appear to shield the vital organs from trauma. The "ear" of the heart appears to protect the aorta and the right coronary cuspid arteries. In the case of the lingula, it appears to protect lung tissue below and also covers part of the heart and blood vessels, including the pulmonary artery anterior cuspid, and left coronary cuspid. The "appendages" are best observed from a transverse (coronal) section of the heart and lung.

Summary and Conclusions

There are dozens of examples that demonstrate the relationship of human structure to function. We have considered only a few of them. As a reader, you will doubt- less discover many more. For example, in the mutualistic relationship between *Escherichia coli* bacteria and our intestinal tract, the bacteria are partially responsible for the remarkable function among the villi, microvilli and folds of the intestinal wall. Each of these structures effectively lengthens the intestine by providing increased surface area for absorption of essential nutrients. Table 4.1 portrays 18 basic designed structures from six body systems. Hundreds of other examples of form fitting function in the human body could be given. Various designs of structures in the skeletal, skin, respiratory, circulatory, immune, and digestive systems attest to their remarkable, purposeful functioning. Could any of these precise interrelationships have originated by nondirected, physiochemical forces? Would chemical molecules fall together by chance to produce such intricate systems? To reproduce order out of disorder, code from non-code, and program from non-program, life forms need energy that is precisely directed. Only when energy is directed and transformed can a chaos of chemicals produce order. Codes and designs with precise functioning presuppose an intelligent cause behind them. We believe that the wonderful patterns, adaptations, and precise functions of the human body parts are clear indications of an Intelligent Designer.

Automotive engineers try each year to improve new automobiles. The engineers' blueprints take sport, durability, cost effectiveness, safety, and personal preferences into consideration. If you ever study the features of cars like Cadillac, Lincoln, Mercedes Benz, Volvo, or BMW, you will see similar trademark signs. Each company has its own impressive logo or a quality "landmark."

Although cars keep improving each year by running faster, becoming more energy efficient, looking sportier, containing more electronic gadgets, and in general, getting better, they will never approach what is really the "ultimate machine," the human body (Kaufmann, 1995). It functions with such precision that human physiologists will never fully understand its superlative qualities. Each year it seems, anatomists, physiologists, and biochemists discover

new mechanisms that more fully explain its construction.

One highly praised anatomy and physiology textbook states, *"Anatomical structures seem **designed** (emphasis ours) to perform specific functions"* (Thibodeau and Patton, 1997). Each body structure has a particular size, shape, form or position in the body related directly to its ability to perform a unique and specialized activity. The most logical explanation for such a high correlation between structure and

Table 4.1. Relationship of Designed Structure to Purposeful Function in Body Systems. Modified from Kaufmann (1995, p. 242) with permission.

System	Design Structure	Purposeful Function
Circulatory	Atria are small and thin walled	Allow for pumping of blood into ventricles with less pressure
	Left ventricle is larger and thicker than the right ventricle	Left ventricle must pump blood to the entire body, while the right ventricle only pumps it to the lungs
	Heart is surrounded by double bound sac filled with serous fluids	Double walled sac acts as a lubricating bag around expanding and contracting heart to reduce friction
	Red blood cells lose their nucleus and become biconcave discs	Allows more room for more hemoglobin to pick up oxygen
	Capillary diameters are somewhat larger than diameters of red blood cells	Membrane of red blood cells are in contact with capillary wall which facilitates exchange of gases
Digestive	Small intestine has finger-like processes (villi) that have microscopic finger-like projections on them	Greatly increases the surface area of small intestine
	Small intestine has larger and thicker wall than large intestine	Small intestine carries out greater moving, mixing, and absorption than the large intestine
	Stomach is lined with hills (rugae) of mucous cells that release mucus	Rugae increase surface area for absorption; protective mucus prevents gastric acids from digesting stomach's own wall
Immune	Spleen has elastic partitions and a smooth muscle covering	Can contract to add a small amount of blood to circulation when there is blood loss

function is a planned origin.

In the same way a specific key fits into a door lock to unlock the door, many body parts are fitted to perform a specific function. In the same way that an engineer plans a bridge or computer scientist designs a chip to fulfill a specific function, we believe the master bioengineer produced the precisely functioning body parts. We can visualize a skilled cabinetmaker selecting tools from a box to build a fine cabinet. The worker uses a pencil to draw a blueprint, a ruler to measure the length of a door, a hammer to pound a nail, a screwdriver to remove a screw, a saw to cut the lumber, and a latch to attach the door to the cabinet. In all, the carpenter may use a dozen distinctly different tools, each with its own role to play. In order to use these tools effectively to build a cabinet there must be a plan before its construction. In Table 4.1, we present over a dozen planned forms and their functions (Kaufmann, 1995).

According to Behe (1996, p. 196):

...in order to reach a conclusion of design for something that is not an artificial object, or to reach a conclusion for a system composed of a number of artificial objects, there must be an identifiable function of the system.

Table 4.1. Relationship of Designed Structure to Purposeful Function in Various Systems (cont.)

System	Design Structure	Purposeful Function
Respiratory	Right bronchus is wider and shorter than the left bronchus	The right lung is larger than the left, therefore it needs more inflow and outflow of air
	Lungs are made of connective tissues and elastic fibers	Allows for the expansion and contraction of the lungs and this provides a greater exchange of gases with the blood
Skeletal	Bursae sacs are intermingled between bones, tendons, ligaments, and skin	Reduces the friction caused when these parts rub together
	Ligaments can tighten and relax	Allows for pronation of ligaments
	Medial and lateral menisci are seated between the tibial condyles of the knee and the tibia	The menisci prevent side-to-femoral joint rocking of the femur to be transmitted to knee joints where they absorb shock
	Anterior crossing ligament of knee joint is present	Prevents the tibia from sliding when the leg is flexed and prevents one from over-extending the calf
	Posterior crossing ligament of knee joint is present	Prevents a backwards displacement of the tibia and allows the ligaments to lock when one is standing
Skin	There are no oil glands on the palms of the hands or soles of the feet	If these areas were oily it could cause one to lose his grip or footing

Harvey and other qualified anatomists convincingly tell us that the heart and lungs have clear functions that are vital for life. Here we can apply again Behe's principle of design detection:

In considering design and the function of the system we must look at the one that requires the greatest amount of the system's internal complexity. We can then judge how well the parts fit the function.

Behe, 1996, p. 196

The anatomy of the heart and the lungs are wonderfully constructed to fulfill their role of transporting oxygen, vital nutrients, and waste products to their required destinations. The tasks that the heart and lung fulfill are best defined by their internal logic of chambers, valves, and complex, branching blood vessels. The most plausible explanation for this high correlation between structure and function of body parts is that it was planned by an Intelligent Cause!

Chapter Five
Homeostasis: The Body in Balance

It is the body that is the hero, not science, not antibiotics, not machines or new devices. The task of the physician today is what it has always been, to help the body do what it has learned so well to do on its own during its unending struggle for survival—to heal itself.

Ronald. J. Glasser, M.D. (1976, cover jacket)

Physicians have long recognized that the body must be in chemical balance to stay healthy. If one chemical variable in the human body gets out of balance, then disorder, disease, and even death may result. Even in ancient Roman times, physicians tried to assist the self-corrective nature of the human body to overcome its disorder. *Vis medicatrix naturae* is the Latin word for the recuperative or self-corrective nature of the human body. Ronald Glasser (1976, p. 1) emphatically says, "The body is the hero!"

Although ancient philosophers had some idea of the balanced condition in the body, it was Claude Bernard, a French physiologist of the nineteenth century, who developed a clear idea of chemical balance. Bernard referred to this body balance as the "constant internal milieu." He recognized the power of many animals to maintain a relatively constant condition in their internal chemical environment even when the external chemical surroundings change. A pond dwelling *Hydra* is powerless to affect the temperature of the fluid that soaks its cells, but the body can maintain its "internal pond" temperature at 37°C, according to Bernard. His initial ideas laid the groundwork for the concept that today is called homeostasis (Campbell, 1996, p. 791).

Homeostasis: Interdependent Processes that Regulate the Body

The word **homeostasis** comes from two Greek terms, *homeo* (alike or the same) and *stasis* (standing or remaining). Thus the word means remaining the same. It is applied to the internal chemical conditions of living things. An American physiologist, Walter B. Cannon (1871-1945), coined the term homeostasis to describe Bernard's original idea of balance and constancy of chemical conditions in living organisms. Cannon's idea was first published in a book, *The Wisdom of the Body* in 1932 (Van de Graff and Fox, 1995).

The "wisdom of the body" points to a wise Creator who lovingly provides and sustains life through these essential homeostatic mechanisms. Cannon viewed the body as a community that consciously seeks out the most favorable conditions for itself. It corrects imbalances in fluids and salts, mobilizes to heal itself, and deploys resources on demand. Homeostasis includes not only the constancy of chemical conditions inside an organism, but also the feedback mechanisms that maintain that constancy (Van de Graff and Fox, 1995, p. 17).

Homeostasis applies to all organisms, but this discussion will be limited to some phases of homeostasis in the human body. The mechanism of homeostasis can be observed in most of the 11 body systems. The movement of a pendulum is an analogy for homeostasis. The moving pendulum represents the fluctuation of a physiological variable, such as body temperature or blood pressure, around an ideal value.

Disturbance in either direction could move the variable into the abnormal range, or into a harmful state. If the disturbance is so extreme that it goes "off the scale" or outside the range of what is tolerable, disease and consequently death can result. Body regulation of these systems is controlled by the nervous and endocrine systems. The Divine Designer has obviously set limits or boundaries on changes in the human body.

Thermostat Wars and Control of Body Temperature

Did you ever wonder why your body temperature remains about 98.6°F whether you are exposed to freezing temperatures in the winter or to temperatures above 100°F in the summer? Man is said to be homeothermic which means that he maintains a constant temperature regardless of the temperature of his environment. Mammals and birds are **homeothermic**. (In contrast, fishes, reptiles, and amphibians are **poikilothermic**; that is, they are of the same temperature as the environmental temperature endured by homeothermic animals.) In homeothermy, the hypothalamus is responsible for our fairly stable temperature.

There are two main ways man's body temperature is regulated to remain constant. First, the amount of heat lost or removed from the body is controlled. More heat is produced by metabolism than is needed (unless the environment is quite cold), so excess heat must be removed. Sweating is one method of removing the heat. When you exercise vigorously or the air around you is warm, sweating increases. When you are still or the surrounding air is cool, sweating decreases. Evaporation of sweat removes excess heat.

The body also removes excess heat from the blood in the skin. When you get hot, your skin gets red. This is because blood vessels in the skin enlarge, permitting more blood to flow near the surface of the skin. The heat in the blood warms the skin and this heat is transferred from the skin into the air. When you are cold, your skin turns pale because the blood vessels contract keeping the blood deeper in the body so that less heat is lost to the air. If there is much pigment in the skin, the red color or the paleness will not be as evident as it will be in lightly pigmented people.

The second way body temperature is regulated is by controlling the amount of heat produced. When muscles contract, because of chilling, this causes shivering and chattering of the teeth. You might walk about and rub your hands in order to generate heat. These processes generate heat. When the air is warm, as on a summer day, your muscles relax so that less heat is generated. That is one reason you may sometimes feel physically lazy during hot weather.

The heat-regulating center in the brain is in the hypothalamus of the diencephalon. Nerve endings in the skin are sensitive to temperature changes. When stimulated, these nerve endings send messages to the heat center. Neural messages are also sent to the sweat glands, blood vessels, and muscles, directing them as necessary to keep the temperature of the body constant. Among the animals there are variations in the ways body temperature is regulated.

Regulation of Heart Rate and the Circulatory System

The heart must meet the regular metabolic need of the body and the speed of beating must also vary according to the demand on the body. During exercise or excitement, the heart beats faster and it beats more slowly when you sleep. Blood pressure also varies throughout the day.

The center for the control of circulation is located in the medulla of the brain stem. Two pairs of nerves travel from the medulla, down the spinal cord, and out to a small neural structure called the pacemaker, located in the right atrium of the heart. One pair of the nerves speeds up the heart rate, and the other pair, called the vagus nerves, slows it down. Because they belong to the autonomic nervous system, they cannot be directly controlled by the conscious mind.

During exercise, there is an increase of blood being returned to the heart through the veins. This increase of blood in the large veins and right atrium causes impulses to be sent to the medulla. The medulla then sends impulses to the pacemaker that speed up the heart rate. If this were to continue unchecked, the heart would beat too fast. There is another mechanism, however, that prevents the heart rate from exceeding its safe limits. The increase of blood pressure in the aorta and in the carotid artery in the neck causes impulses to be sent to the medulla, which then slows the pacemaker. The two processes, heartbeat increase and heartbeat decrease, are kept in balance so that the heart beats at the right speed to meet the demands of any particular exercise.

The diameter of the blood vessels is also controlled automatically. During exercise, the blood vessels in the skin, heart, lungs, and muscles become larger. The dilated vessels in the dermis of the skin provide for eliminating heat. The heart, lungs, and muscles need an increased amount of blood to deliver nutrients to the working cells and to remove wastes. When this happens, blood vessels in the digestive tract decrease in size. This constriction actually slows digestion. If all the blood vessels were enlarged at the same time, there would not be enough blood to fill them and metabolic function would decline. When a person is resting, the blood vessels of the skin and other organs constrict somewhat. When food is being digested, the vessels in the stomach and intestines enlarge to allow the chemical digestion of the food to occur. Capillaries intertwined in the intestinal villi then carry the nutrients to the liver for processing and for delivery to the working muscles and organs via the arteries. Eventually these nutrients are converted into energy-rich molecules, like glucose, to be oxidized for the body's needs.

Glucose levels are very important in homeostatic control. For example, the adrenal glands, and the alpha and beta cells in the pancreas, control the amount of glucose in the blood, which in turn affects the balance between hunger and satiety. There are many other complex relationships between glands and body regulation (Tortora, 1994; Tortora and Grabowski, 1996).

These are balanced interactions necessary to maintain homeostasis in the body. These interdependent biochemical connections are additional examples of what Behe calls irreducibly complex systems in the body. If you take away any one of these vital internal control mechanisms, the body will be unable to maintain, or regulate its steady state. These evidences support the belief that an all-wise Creator produced the body fully developed.

Balance and Design in the Body's Urinary System

Further evidence of design is seen in the urinary system. The kidneys appear to have the type of organization which is a result of intelligent planning. The kidney is one of the unsung heroes of the body. These bean-shaped structures function primarily to regulate the body's extracellular fluid. This function is accomplished in part through the formation of urine, a product of filtered and modified blood plasma.

In the process of urine formation, the kidneys regulate (1) the blood plasma volume and, in turn blood pressure; (2) the concentration of waste products in the blood; (3) the concentration of electrolytes (Na^+, K^+, HCO_3^-, etc. in the plasma; and (4) the plasma pH (Van de Graff and Fox, 1995). Because the design inference is most evident in complex and interdependent feedback systems, we will focus on the last two functions of the kidney. In the next sec-

tion, note the interacting components necessary to keep the body in balance.

The Design Inference and Water Balance

The term **balance** suggests a state of equilibrium, and in the case of water and electrolytes, it means that the quantities entering the body are equal to the quantities leaving it. Maintaining such a balance requires mechanisms to ensure that lost water and electrolytes will be replaced and that any excesses will be expelled. In fact, an in-depth study would show it is an **optimal balance**, one that minimizes metabolic cost to the organ and system; while at the same time maximizes benefit (fluids, electrolytes and energy) to the entire body. This optimal balance is a result of nearly perfect efficiency of the body systems. As a result, the quantities of water, electrolytes and H^+ ions within the body are relatively stable at all times.

It is important to remember that water balance and electrolyte balance are interdependent, because the electrolytes are dissolved in the water of the body fluids. Consequently, anything that alters the concentrations of the electrolytes will necessarily alter the concentration of the water by adding solutes to it or by removing solutes from it. Likewise, anything that changes the concentration of the water will change the concentrations of the electrolytes by making them either more concentrated or more diluted.

Water and electrolytes are not uniformly distributed throughout the tissues. Instead, they occur in regions, or compartments, that contain fluids of varying compositions. The movement of water and electrolytes between these compartments is regulated, so that their distribution remains stable. Water balance exists when the total intake of water is equal to the total loss of water.

The primary regulator of water output is urine production. The volume of water excreted in the urine is regulated mainly by activity in the distal convoluted tubules and collecting ducts of the nephron. The epithelial linings of these segments of the renal tubule remain relatively impermeable to water unless vasopressin, or antidiuretic hormone (ADH), is present. The action of ADH causes water to be reabsorbed in these segments and, thus, to be conserved. In the absence of ADH, less water is reabsorbed and the urine volume increases (Hole, 1995).

The Design Inference and Electrolyte Balance

An electrolyte balance exists when the quantities of the various electrolytes gained by the body are equal to those lost. The electrolytes (substances that release ions in water) of greatest importance to cellular functions are those that release the ions of sodium, potassium, calcium, magnesium, chloride, sulfate, phosphate, and bicarbonate. These substances are obtained primarily from foods, but they may also occur in drinking water and other beverages. In addition, some electrolytes form as byproducts of various metabolic reactions.

Ordinarily, a person obtains sufficient electrolytes by responding to hunger and thirst. When there is a severe electrolyte deficiency, however, a person may experience a strong desire to eat salty foods.

The concentrations of positively charged ions, such as sodium (Na^+), potassium (K^+), calcium (Ca^{+2}) and hydronium (H^+ or H_3O^+), are particularly important. Certain concentrations of these ions, for example, are necessary for the conduction of nerve impulses, contraction of muscle fibers, and maintenance of cell membrane permeability. Sodium ions account for nearly 90 percent of the positively charged ions in extracellular fluids (Figure 5.1). The primary mechanism regulating these ions involves the kidneys and the hormone aldosterone. This hormone is secreted by the adrenal cortex. Its presence causes an increase in sodium reabsorption in the distal convoluted tubules and the collecting ducts of the renal tubules.

The sodium/potassium ion balance is regulated by aldosterone from the adrenal cortex. Aldosterone also functions in regulating potassium. In fact, the most important stimulus for aldosterone secretion is a rising potassium ion concentration, which seems to stimulate the cells of the adrenal cortex directly. This hormone enhances the reabsorption of sodium ions and at the same time causes the secretion of potassium ions into the renal filtrate.

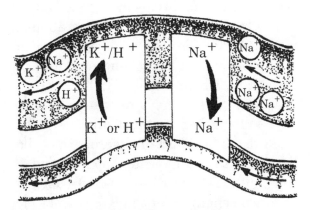

Figure 5.1. Movement of ions through a peritubular capillary and distal renal tubule.

The concentration of calcium ions in extracellular fluids is regulated mainly by the parathyroid glands. Whenever the calcium ion concentration drops below normal, these glands are stimulated directly, and they secrete parathyroid hormone in response. Parathyroid hormone causes the concentrations of calcium and phosphate ions in the extracellular fluids to increase (Hole, 1995; Hole, Shier, Butler, and Lewis, 1996).

The Design Inference and pH Balance

One of the most important functions of the kidneys is to maintain the balance of the acids, bases, and salts in the blood. Body fluids must remain at the optimal pH (relative measure of acidity or alkalinity in solution) level in order to resist extreme acidosis or alkalosis. During exercise, more acids are produced. There are two ways of correcting the excess of acid, or hydrogen ions. One is by buffers, which are chemicals that neutralize strong acids or strong bases in the bloodstream.

The second means of controlling acidity is the respiratory system that helps maintain constant conditions in your body. One of the by-products of aerobic metabolism is carbon dioxide. It forms carbonic acid in the blood that is broken down into bicarbonate, which gives off CO_2 to the exposed air. The kidneys remove excess acids resulting from metabolism, chiefly lactic acid, a by-product of anaerobic metabolism and the same acid that is found in sour milk. The liver has a wondrously intricate mechanism for changing excess lactic acid back into glucose, which can be used for energy.

The Respiratory Center

The nervous center that controls breathing is in the medulla of the brain stem. It is called the **respiratory center.** The medulla can be stimulated by the amount of carbon dioxide in the blood. If too much carbon dioxide accumulates, the medulla causes deeper and faster breathing to remove CO_2 rapidly. (Carbon dioxide in the blood combines with H_2O to form H_2CO_3, carbonic acid.) This happens during exercise. A low carbon dioxide level, on the other hand, causes the medulla to signal slower and shallower breathing.

This feedback mechanism causes you to breathe without effort, from birth to death. The automatic control of breathing keeps the composition of the blood stable and constant. It provides a steady supply of oxygen to the working cells and helps maintain the volume and composition of the blood.

Sometimes swimmers take several deep breaths before diving so they can stay under water longer. By removing large amounts of carbon dioxide, they eliminate the need to breathe as often. This can be dangerous, however, since dizziness or even unconsciousness can result.

The respiratory center in the brain stem also helps regulate hydrogen ion (H^+) concentrations in the body fluids by controlling the rate and depth of breathing. Specifically, if the cells increase their production of carbon dioxide, as occurs during periods of physical exercise, the production of carbonic acid increases. As the carbonic acid dissociates, the concentration of hydrogen ions increases, and the pH of the fluids tends to drop. Such an increasing concentration of carbon dioxide and the consequent increase in hydrogen ion concentration in the plasma stimulates chemosensitive areas within the respiratory center.

In response, the respiratory center causes the depth and rate of breathing to increase, so that a greater amount of carbon dioxide is excreted through the lungs. This loss of carbon dioxide is accompanied by a drop in the hydrogen ion concentration in the body fluids, because the released carbon dioxide comes from carbonic acid:

$$H_2CO_3 \longrightarrow CO_2 + H_2O$$

Conversely, if the body cells are less active, the concentrations of carbon dioxide and hydrogen ions in the plasma remain relatively low. As a consequence, the breathing rate and depth are decreased.

The Kidneys

Another means of controlling acid-base balance is the removal of excess acids from the body. This removal takes place in the nephron by excreting hydrogen ions from the kidneys. These ions are secreted into the urine by the epithelial cells that line certain segments of the renal tubules. By controlling the amount of electrolytes, acids, and water removed from the blood, the kidneys regulate the quantity and composition of the blood at an optimal functional level (Marieb, 1994).

The various regulators of hydrogen ion concentration operate at different rates. Acid-base buffers, for example, function rapidly and can convert strong acids or bases into weak acids or bases almost immediately. For this reason, chemical buffer systems sometimes are called the body's first line of defense against shifts in the pH.

Physiological buffer systems, such as the respiratory and renal mechanisms, function more slowly, and constitute secondary defenses. The respiratory mechanism may require several minutes to begin resisting a change in the pH, and the renal mechanisms may require one to three days to regulate a changing hydrogen ion concentration (Hole, Shier, Butler, and Lewis, 1996).

Acid-Base Balance

One reason to believe the design inference is that the delicate and complex balance would have had to be planned from the beginning, or nature has to overcome tremendous obstacles that are nearly impossible to evolve the level of complexity necessary to achieve homeostasis through gradual physiochemical mechanisms (i.e. mutations and natural selection). The creeping towards balance is very improbable.

Figure 5.2 illustrates that the pH of blood plasma is maintained within a narrow range of values through the function of the kidneys (in conjunction with the lungs). The kidneys regulate the bicarbonate concentration and the lungs regulate the carbon dioxide concentration. The kidneys regulate acid base balance by the secretion of hydrogen (H^+) ions into the tubules and the reabsorption of bicarbonate (HCO_3^-). The amounts of H^+ secreted and HCO_3^- reabsorbed are interdependent; that is the reabsorption of HCO_3^- occurs as a result of the filtration and secretion of H^+. The amount of H^+ filtered and secreted varies depending on the pH of the

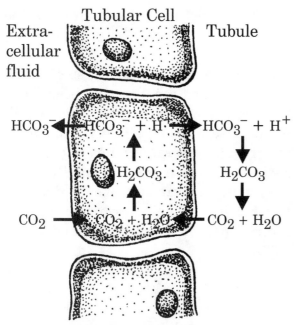

Figure 5.2. Kidney tubule cells showing maintenance of acid-base balance.

body fluid. In **acidosis**, the ratio of CO_2 to HCO_3^- in the extracellular fluid is increased because of the production of CO_2 or increased H^+ formation from metabolites. The renal response is to balance the extra H^+ ions. The net result is that H^+ ions are excreted from the body and HCO_3^- ions are retained.

The kidneys (Figure 5.2) regulate acid base balance by the secretion of H^+ ions into the tubules and the reabsorption of bicarbonate (HCO_3^-). The slightest imbalance of body fluids with H^+ ions leads to acidosis. In acidosis, the ratio of CO_2 to HCO_3^- increases in the extracellular fluid because of the production of CO_2 or increased H^+ formation from metabolites. As you will soon see, the slightest imbalance may be deleterious, if not disastrous, in consequences if imbalance were sustained. On the other hand, the slightest imbalance of body fluids with OH^- ions lead to alkalosis. In alkalosis, the ratio of HCO_3^- to CO_2 increases in the extra-cellular fluid and the

pH rises. Some changes in body fluid pH cause marked alterations in the rates of chemical reaction in cells and in overall body function. Acidosis may lead to death as a result of coma and alkalosis may lead to death as a result of tetany or convulsions.

Figure 5.3 illustrates the wisdom, skill, and care of the Creator for the body's acid-base regulation by providing such a precise balance in the "dust of the earth"(Genesis 2:7). Ordinarily, the hydrogen ion concentration of body fluids is maintained within a very narrow pH range by the actions of chemical and physiological buffer systems. However, disease conditions may disturb the normal acid-base balance and produce serious consequences. For example, the pH of arterial blood is normally about 7.4 (7.35–7.45), and if this value drops below 7.35, the person is said to have acidosis. If the pH rises above 7.45, the condition is called alkalosis. Such shifts in the pH of body fluids may be life threatening, and in fact, a person usually does not survive if the pH drops to (6.8 or rises to 8.0) for more than a few hours.

Acidosis is caused by an accumulation of acid or a loss of base, resulting in an increase of the hydrogen ion concentration of body fluids. Conversely, alkalosis is caused by a loss of acid or an accumulation of base, accompanied by a decrease in hydrogen ion concentration. Two types of acidosis are **respiratory acidosis** and **metabolic acidosis.** Respiratory acidosis is caused by conditions that hinder pulmonary ventilation, such as injuries to the respiratory center in the brain stem, obstructions in the air passages, or diseases that reduce the surface area of the respiratory membrane and decrease the volume

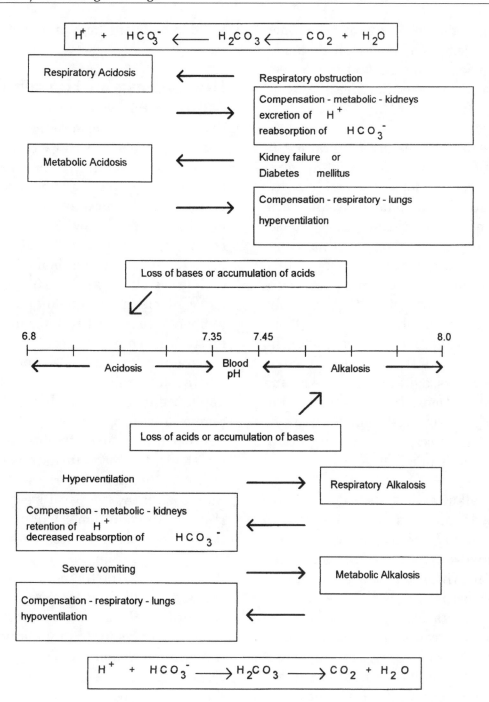

Figure 5.3. Acid Base Balance Mechanisms in the Human Body.

of gas exchanged in the lungs. The symptoms of respiratory acidosis include labored breathing, drowsiness, disorientation, cyanosis, and depression of the central nervous system, characterized by stupor.

Metabolic acidosis occurs when nonres-

piratory acids accumulate or bases are lost. These bases are depleted by metabolic processes in renal disease. The excretion of acids produced by protein degradation is increased while vomiting of the upper intestine contents are decreased, thereby many alkaline contents lost. Sometimes in diabetes mellitus fatty acids are converted into ketone bodies leading to metabolic acidosis. The symptoms of metabolic acidosis are nausea, vomiting, and rapid, deep breathing. Prolonged diarrhea may continue.

Two types of alkalosis are respiratory alkalosis and metabolic alkalosis. Respiratory alkalosis develops as a result of hyperventilation accompanied by an excessive loss of carbon dioxide and a consequent decrease in carbonic acid, and hydrogen ion concentration in body fluids. Hyperventilation most commonly occurs during periods of anxiety, although it may also accompany fever or the toxic effects of certain drugs, such as aspirin. The symptoms of respiratory alkalosis include lightheadedness, agitation, dizziness, and tingling sensations. In severe cases, impulses may be triggered on motor neurons and skeletal muscles may undergo tetanic contractions in response.

Metabolic alkalosis results from excessive loss of hydrogen ions or from a gain in bases. This condition may occur following the removal of gastric juice, prolonged vomiting in which only the stomach contents are lost, or ingestion of excessive amounts of antacids, such as sodium bicarbonate. The symptoms of metabolic alkalosis include a decreased rate and depth of breathing, irritability, muscular weakness, and decreased intestinal motility.

In summary, the consequences for our bodies when water, electrolytes, and/or pH get out of balance is disease, permanent disorder, and sometimes even death. The consequences may be severe when the pendulum of balance is off one way or the other. Natural selection, mutation, and chance physiochemical processes hardly could explain the body in balance.

Homeostasis in Household Appliances

Cybernetics is a term applied to the science of controls in both living and nonliving systems. Technically, homeostasis applies to living systems only, but there are systems like homeostasis designed by human engineers. The cybernetics of non-living systems deals with feedback mechanisms as in the case of home heating and cooling systems. Examples of homeostatic-like controls in household appliances and temperature control systems include thermostats in your refrigerators, air conditioners and furnaces. Each of these thermostats controls the temperature of an appliance or the entire home. When the temperature drops below the desired level, the thermostat turns the furnace or heating element on. The furnace is turned off when the temperature rises above a certain level. If the temperature continues to rise, the air conditioner will be turned on. It is shut off when the temperature drops to the level set on the thermostat. With these systems, the temperature of a house can be kept within a narrow optimal range year around. Both heat production and its removal are regulated by a kind of "electromechanical homeostasis."

Parallels can be drawn between the control of the temperature of a house and temperature control in homeothermal animals. In both cases, there is a *stimulus*, a *receptor*, a *control center,* and an *effector*. The receptor detects fluctuations in some variable of the animal's internal environment such as a change in temperature (stimulus). The control center processes information it receives from the receptor and directs an appropriate response by the effector.

When the temperature falls below a **set point**, the thermostat switches on the heater (the effector). When the thermometer detects a temperature near the set point, the thermostat switches off the heater. This type of control circuit is called **negative feedback** system because a change in the variable being monitored (heat) triggers a response that counteracts the excess of that same variable—heat. Negative feedback mechanisms prevent small changes from becoming too large. Most homeostatic mechanisms involving diverse body systems in humans operate on the principle of negative feedback (Campbell, 1996; Marieb, 1994).

This similarity may leave the impression that the systems are alike and animals are entirely mechanistic. Closer inspection, however, reveals a vast difference, because homeostasis is far more complex than temperature controls in household appliances. The thermostat in the refrigerator or house is stimulated only by air temperature. In living systems, the rate of metabolism, which is controlled by hormones, affects temperature independently of the environment. Also, fever results when the temperature setting is raised as a result of infection of pathogens (germs).

Homeostasis is a phenomenon in living systems and is vastly more involved than control systems in appliances invented by man. The complex system of checks and balances has been invented by a superior cause, the Creator. The Inventor is superior to the invented. The advanced thinking that went into His systems far exceeds anything man has been able to develop.

How marvelous and wise is the Bio-engineer of the human body!

Homeostasis: The Creator's Common Blueprint for Constancy

We have explored how design principles operate in the excretory system and how homeostasis, order, and the boundaries of the urinary system serve to protect the human body. Homeostasis is an organized and irreducible property of order that maintains the body's delicate balance through feedback (usually negative). This delicate balance is the mark of a creative genius—an intelligent Designer if you will.

It is the constancy of homeostasis that is the condition of free and independent life. All the mechanisms of life, no matter how varied, have the purpose to keep the conditions of life constant in the internal environment. God who is discerned in nature must be doing much of his work in the *milieu interieur*. It is homeostasis, the dependability and steadiness of the internal environment, that keeps us alive. The structure of the kidneys is based on their critical role in maintaining homeostasis (Nuland, 1997).

Indeed the Creator put wisdom in the "inward parts" (Job 38:36). This wisdom is evident in the process of homeostasis: balance, order, regulation, and feedback. It is the Creator who has given understanding to the mind of man as he has discovered the laws that the Creator set in motion in the human body. As we study many different body systems, their glands and hormones, much of the body's design is to keep this marvelous, delicate balance that maintains life.

Homeostasis is a universal characteristic of all living things. We have studied homeostatic control in humans, but these systems are also found throughout the animal kingdom. There are many variations in the way this is done, but the basic principle is the same. Most human homeostatic control systems involve three or more components: the endocrine, nervous, and one other system, coupled with many biochemical reactions. Homeostatic controls are irreducibly complex in nature. This irreducible complexity in humans involves the intricate living controls, whereas nonliving control mechanisms are quite simplistic by comparison. It is most unlikely that such an intricate and delicate balance would have developed by chance from genetic mutations that are largely harmful!

Historically the first scientist to use the metaphor of wisdom to characterize the body in balance was Ernest Starling (co-discoverer of hormones) in 1923. His ideas stemmed from his observation that the body seemingly had an intuitive integration of its diverse faculties. In his delivery of the prestigious Havreian Oration of the Royal College of Physicians in Great Britain, he spoke about the regulation of body processes, their adaptability, and the contribution of hormones toward integrating them into a single system. For his epigraph, Starling chose a verse from the Book of Job:

Who hath put wisdom in the inward parts? or who hath given understanding to the heart?

Job 38:36

Starling associated the "body in balance" with the wisdom found in the inward parts designed by the Creator. He spoke about this coordinated communication among cells like the lung and kidney, and developed mechanisms describing regulatory processes, like the acid-base balance (Figures 5.2, 5.3). Nine years after hearing this famous oration, Walter Cannon coined the term **homeostasis** in his book, *The Wisdom of the Body*, and built upon Starling's theory. In turn, Starling further advanced (Evans, 1949) the theory of homeostasis when he described this as a condition of uniformity that results from the adjustment of living things to changes in their environment. He described detailed physiological mechanisms for this coordinated regulatory balance (Evans, 1949). "All the mechanisms of life, no matter how varied they are, have only one object, to keep the conditions of life constant in the internal environment" (Nuland, 1997, p. *xvii*). God must be doing much of His work in the *milieu interier* (Bernard's term), or inward parts (Starling's term) to demonstrate that He is the Intelligent Designer.

Chapter Six
Multifaceted Systems in the Human Body

To be a member is to have neither life, being, nor movement, except through the spirit of the body, and for the body.

Blaise Pascal (Brand and Yancey, 1980, p. 21)

"A human being is more than the sum of its parts" is the major theme of a *Life* magazine article by Alex Tsiaras (1997). In the beautifully illustrated photo essay, "A Fantastic Voyage through the Human Body," the dynamic nature and unique properties of the eleven organ body systems are discussed and pictured. Each multifaceted system illustrates the elaborate design and the interdependence of the many parts in the human body. Any absent part in one of the body's many "adaptational packages" could lead to dysfunction of that system and, sometimes, even lead to death.

In this chapter, we have selected several familiar examples from the human body to illustrate the design argument of irreducible intricacy. There are many "molecular and cell teams" that illustrate this fea-

ture. A question to ponder as you study these particular examples is: Are the mechanisms proposed in the "Blind Watchmaker" argument (Dawkins, 1996a) sufficient to explain this remarkable order and complexity in human anatomy and physiology? Will time and chance produce such wondrous and complex systems by evolution?

The Hands and Their Movement
The movement of the hand and fingers of a concert pianist is an awesome sight. The necessity of coordination, timing, and order to play Beethoven's "Fifth Concerto" or Bach's "Jesu, Joy of Man's Desiring" is a feat that is not accomplished by chance.

There is marvelous skill not only in playing the music, but in the seventy (35 in

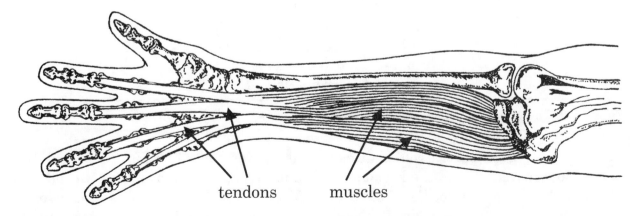

tendons   muscles

Figure 6.1. The hand and its movement by muscles and tendons.

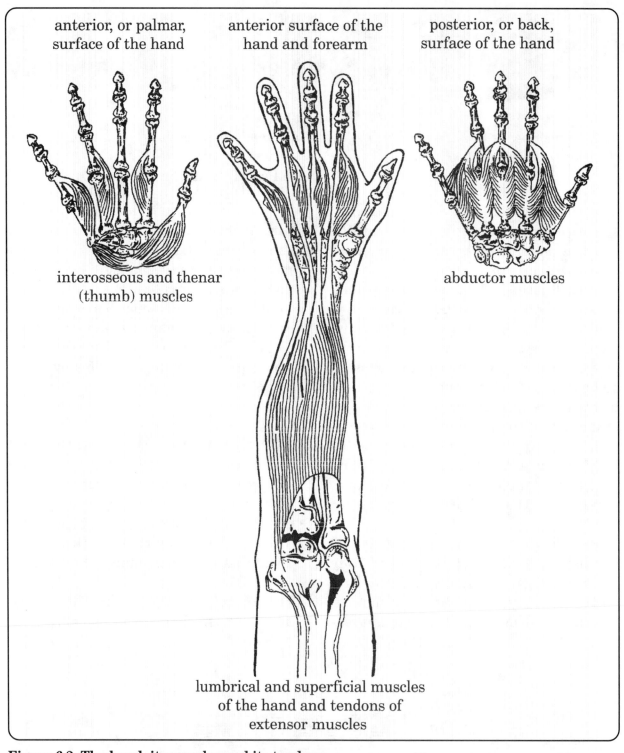

Figure 6.2. The hand, its muscles and its tendons.

each hand) separate muscles contributing to the hand movements on the keyboard. The hand has been described as the most sophisticated "tool" in the body. It looks like it was crafted for maximum dexterity and strength in movement (Brand and Yancey, 1980). The hand is capable of 58 distinct movements. These movements allow for dexterity and power for a diversity of actions ranging from piano playing and threading a needle to holding a jackhammer. This amazing diversity of functions is accomplished with the help of muscles in the forearm and wrist. The fingers have no muscles in themselves; the tendons transfer force from muscles in the forearm and palm.

Most hand movements are promoted by the forearm muscles (Figures 6.1 and 6.2). These movements are assisted and made more precise by a number of small *intrinsic* muscles in the hand. These hand muscles include the four *lumbrical* muscles that lie between the metacarpals (visible on the palm surface), and the seven *interosseous* muscles that lie deep beneath the skin and inferior to the lumbricals. Together, the lumbricals and interossei flex the knuckles. They also extend, adduct, and abduct the fingers. There are also four *thenar* muscles that act exclusively to circumduct and oppose the thumb.

Orthopedic surgeons could write many manuals suggesting various ways to repair hands that have been injured. Yet, there has never been a surgical technique that succeeded in improving the movement in a normal healthy hand. It frequently takes over a dozen muscles and tendons working together with the opposable thumb to accomplish one movement. No wonder Sir Isaac Newton was convinced of intelligent action in creation from his study of the thumb alone (Brand and Yancey, 1980)!

The Dynamic Digestive System

The news media publish many articles about microbes that cause disease but few articles about microbes that are useful. In fact, "microbe phobia" permeates our society (Williams and Gillen, 1991). Actually, only about 5% of all bacteria are pathogenic. Many bacteria are beneficial and some are even essential for human life. This relationship between microbe and man might be called an "adaptational package."

An **adaptational package** is a biological relationship in which the whole is functionally more than the sum of its individual components. *Escherichia coli* and related enteric colon bacteria are the predominant microbes in the lower intestine, and they constitute about 75% of all the living bacteria in feces. Biologists refer to this cooperation between *E. coli* and the colon as mutualism, a relationship where both species benefit from living together.

Bacteria are not uniformly distributed throughout the length of the intestine. There are differences in the numbers and kinds of microorganisms at different levels. The empty stomach is nearly sterile, and there are few organisms in the duodenum and upper jejunum. These low bacterial numbers are probably caused by the acid secretion of the stomach.

The lower levels of the small intestines that are alkaline become progressively richer in bacteria, and in the adult large intestine the number of microorganisms reaches its maximum. Besides *E. coli*, there are 34 intestinal genera of aerobic and anaerobic bacteria. Our diet markedly influences the relative abundance of these bacteria in the feces.

The majority of microorganisms in the intestines under normal conditions do no harm. Indeed, the intestinal bacteria con-

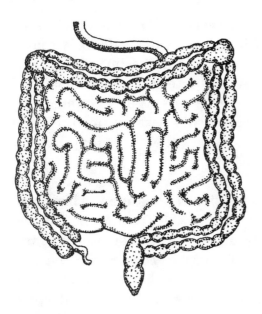 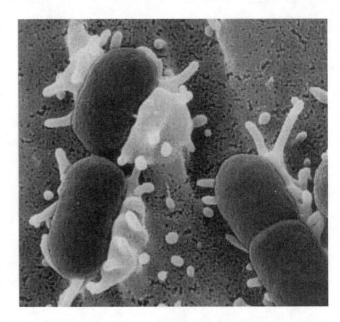

Figure 6.3. The small and large intestines (left), and the *Escherichia coli* bacteria (right) that inhabit the large intestine.

tribute to the general well-being of both microbes and people by synthesizing a number of the vitamins essential for good nutrition and breaking down various macronutrients. The human body cannot synthesize niacin or niacinaminde to make nicotinamide adenine dinucleotide (NAD), which is necessary for energy conversion in the cell's mitochondria. Vitamin K, niacin, NAD and B-complex vitamins are formed by bacteria in significant amounts in the colon.

The B vitamins, including niacin, cobalamin (B_{12}) biotin, thiamine, and riboflavin, are necessary for normal energy levels, freedom from fatigue and proper functioning of nerves. A prolonged deficiency in any one of these vitamins may lead to chronic fatigue, and inability to lead a normal life. As for the benefit to *E. coli* (and other enteric bacteria), man's colon provides a stable nutritional home (Dixon, 1976). In summary, both parties benefit.

The mutualistic relationship between our

digestive cells and these helpful microbes is an amazing cooperation and another example of emergent properties in living things. The large intestine contains the highest numbers of resident flora because of the available moisture and nutrients in the colon. If you take away *E. coli* and other bacteria, the function of the digestive system is severely impaired. This cooperation among cells is similar to the interdependent parts of a mousetrap. A summary of benefits is found in Table 6.1.

In addition, intestinal bacteria help to release waste nitrogen, in the form of ammonia that becomes part of amino acids. They modify bile acids and steroids, and they play a role in the metabolism of polyunsaturated fatty acids. Studies in young mammals have demonstrated that bacteria produce enzymes needed to digest lactose in the germ free milk of its mother. Young mammals need glucose in their diet or they succumb to hypoglycemia. Other benefits to mammals having these microbial symbionts include 1) the bacteria's ability to break down food additives such as coloring and flavoring agents in intestines; and 2) enteric bacteria enhance the normal development of villi, functional projections of the small intestine.

Mutualism is therefore evidence for creation. In mutualism there is a marvelous fit between two radically different organisms. Mutualism poses a problem for macroevolution because it requires that the traits of two organisms evolved at the same time. Simultaneous mutations must have occurred during natural selection in two very different organisms.

Already we have pointed out the difficulty of one successful trait evolving from random mutations. In addition, there is a tendency for mutualism to degenerate into parasitism as a result of mutation.

We must ask if there is any evidence for

mutualism having arisen by evolution. There appears to be no record of new mutualism emerging in the last century. It is clear that mutualism could have not simply evolved by chance alone because of the incredibly low probability that both traits would have arisen simultaneously. We must conclude that the most plausible explanation is that human intestinal cells and these bacteria were created to function together.

In summary, the relationship between intestinal bacteria and humans is optimum for survival because of the ability for cooperation within this entire "package of parts." In this case, it is a mutualistic, coordinated system of microbes working together with man's gastrointestinal tract. It is adaptive in that it increases the survival of both human and microbe. In fact, this remarkable unity between these magnificent microbes and our intestines can

Table 6.1. An adaptational package *Escherichia coli* and other enteric bacteria* living mutualistically within the digestive system.

Vitamins and Enzymes	Metabolic Role	Deficiency Diseases
Vitamin K	Needed in synthesis of prothrombin for blood clotting	Hemorrhage in newborn who lack gut bacteria
Vitamin B12 (Cobalamin)	Coenzyme needed in the formation of proteins and nucleic acids	Pernicious anemia, lack of energy
Niacin (Vitamin B3) NAD compounds	Part of NAD+ and NADH coenzymes in energy metabolism	Pellagra, lack of energy, fatigue
Vitamin B-complex (thiamine, biotin, and riboflavin)	Coenzymes needed for carbohydrate, protein, and fat metabolism	Beriberi, pellagra, anemia, fatigue, inflammation and breakdown of skin
Lactase	Breaks down lactose (milk sugar)	Lactose intolerance (cannot digest dairy products)

*Other common enteric bacteria include *Bacteroides, Bifdobacterium, Citrobacter, Enterococcus, Enterobacter, Fusobacterium, Klebsiella, Lactobacillus, Proteus, Peptostreptococcus,* and *Shigella.* There are 34 intestinal genera in all.

be logically derived as having its origin in an Intelligent Designer. This remarkable cooperation is a strong evidence for creation because it defies the laws of probability that two very different organisms can live together in harmony!

Extraordinary Excretory System

All organisms produce wastes. These waste materials must be removed so that the organism is not poisoned by its own metabolic byproducts. In humans, the removal of these wastes is handled by the lymphatic system, the circulatory system, and the excretory system. Many body parts and systems indeed work together to accomplish the work of body "janitoring." The excretory system has extraordinary capabilities to remove nitrogenous wastes from the body through the processes of filtration, reabsorption, and secretion of blood.

Urine is the fluid produced by the kidneys as they remove waste chemicals from the blood. Urine is made up primarily of wa-ter with some electrolytes and organic materials dissolved in it. The concentration of each of these substances varies with a person's health, diet, and degree of activity. By testing the chemical composition of urine, doctors can learn much about the general health of an individual. Urinary tract infections, kidney malfunction, diabetes, and liver disease are just a few of the medical problems that can be diagnosed through urinalysis.

The functional unit in the kidney that produces urine is called a nephron (Figure 6.4). Urine formation results from glomerular filtration, tubular reabsorption, and tubular secretion. Filtration takes place in the glomerulus where water and solutes, smaller than proteins, pass through the glomerular capsule from the blood. Red blood cells are screened and do not enter the nephron in a healthy kidney but remain in the bloodstream or capillary. Tubular reabsorption takes place in the proximal renal tubule, where water, glucose, amino acids, and needed ions are trans-

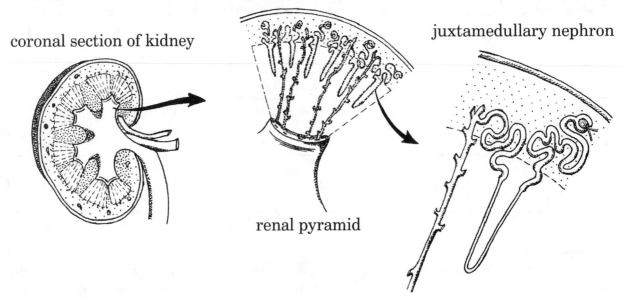

coronal section of kidney

juxtamedullary nephron

renal pyramid

Figure 6.4. The kidney and its basic unit, the nephron.

ported out of the filtrate through cells. They return to the bloodstream in the peritubular capillaries from the nephron.

Tubular secretion is the process in which protons (H^+), ammonia (NH_4^+), potassium (K^+), certain drugs, and creatine are removed from the peritubular capillaries and secreted by renal tubule cells into the filtrate. Eventually some water, urea, and uric acid accumulate in the collecting ducts to form the urine. The urine is removed by way of the ureters, the urinary bladder, and the urethra. After the urinary bladder is partially filled and pressure is exerted, the urine finally exits the body through the urethra channel.

What Makes the Excretory System Extraordinary?

Essentially three observations make the excretory (urinary) system amazing. First, its basic unit, the nephron is a multifaceted system in itself in that it performs functions of filtration, reabsorption, and secretion simultaneously. In addition, the total volume that our kidneys filter in a lifetime, usually without failure, is staggering. Finally, its structure and function precludes its emerging by chance (Figure 6.4).

All body cells are metabolic furnaces, burning glucose and producing waste. Throughout the day, as cells metabolize sugars they produce waste, and release it into the blood. Five to six liters of blood passes through the nephrons at a rate of 1.2 liters per minute. (Tortora and Grabowski, 1996). This translates into all blood filtering through the kidneys about 20 times every hour. The blood passes through 60,000 miles of vessels and picks up waste

from the hardworking cells. The kidneys use water to remove toxins that would otherwise poison the blood. In a normal adult, the excreted rate is about 125 ml/min., or about 180 liters (48 gallons) per day (Tortora and Grabowski, 1996). This amounts to about 17,520 gallons/year or 1,401,600 gallons in an 80-year lifetime. This is an extraordinary amount for a machine to filter. The role of the kidneys in your body is to filter and chemically balance the blood. The kidneys excrete waste products, but also recycle useful elements for body tissues through the circulatory system.

If the kidneys fail, dialysis machines can take over the job. **Dialysis** refers to the process of separating molecules of different sizes using a semipermeable membrane. Dialyzers permit patients to stray far from the hospital room between their three times weekly treatment. From a vein in the arm, blood flows into the unit where it passes by a semipermeable membrane that allows wastes but not blood cells to pass through it. A second filter removes air bubbles before the machine pumps the cleansed blood back into the body.

The dialysis machine can temporarily help people overcome their kidney deficiency. The dialysis machine made by bioengineers substitutes for our kidney for several months. Newer dialysis machines may help some patients with minimal kidney function for periods up to four years. Eventually, however, the patient will want to obtain a kidney transplant. Few people would want to be on dialysis for a lifetime. Yet, everyone would agree that dialysis machines are engineered with preplanned design. How then could anyone say the kidney developed by accident?

Chapter Seven

Clotting Cascades

When Charles Darwin was climbing the rocks of the Galapagos Island, he must have cut his finger occasionally or scraped his knee. Young adventurer that he was, he probably paid no attention to the blood trickling out.

Michael Behe, 1996, p. 77

If Charles Darwin had paid attention to his cuts and bruises on those sharp ocean ledges and knew the biochemistry of blood the way Michael Behe or Dean Kenyon does today, he may have reached a different conclusion about body systems in his *Origin of Species*. Like the multifaceted systems discussed in Chapter 6, the biochemical components involved in blood clotting and wound restoration will function only as an entire unit or not at all.

Clotting involves various chemicals known as coagulation factors. Clotting is a complex process in which coagulation factors activate each other. That is, the first coagulation factor activates the second, the second activates the third, and so on. There are about a dozen factors involved in the two dozen or so reactions that involve clotting and wound restoration (Table 7.1). This is why the blood clotting system is called a **cascade**, a system where one component activates another component. In this chapter, you will explore the complex mechanism behind blood clotting.

As an overview, we describe clotting in three basic stages: **Stage 1.** Formation of prothrombinase (prothrombin activator); **Stage 2.** Conversion of prothrombin (a plasma protein formed by the liver) into the enzyme thrombin, by prothrombinase; and **Stage 3.** Conversion of soluble fibrinogen (another plasma protein formed by the liver) into insoluble fibrin by thrombin. Fibrin forms the network of threads in the clot (see cover). Two cascades may occur during blood clotting, the extrinsic and intrinsic pathways. Prothrombinase is formed as part of two pathways of blood clotting (Figure 7.1). Both pathways will be discussed in depth in this chapter.

Cuts, Scrapes, and Wounds

We all get cuts, scrapes, and wounds from time to time, yet we do not bleed to death because blood contains self-sealing materials that plug the leaks in our vessels and tissues. Blood clotting and wound repair is a complex, intricately woven system consisting of more than two dozen interdependent biochemical reactions. The absence of, or significant defects in, any one of the components causes the system to fail. Blood will not clot at the proper time or at the proper place.

The entire process of blood clotting and wound restoration involves **hemostasis.** Hemostasis refers to the stoppage of bleeding. When blood vessels are damaged,

Table 7.1. Blood Clotting Factors and the Cascading Components (Simplified from Tortora and Grabowski, 1996, p. 567).

Clotting Factor	Pathway	The Cascading Components
I	Common	Fibrinogen
II	Common	Prothrombin
III	Extrinsic	Tissue factor (thromboplastin)
IV	All	Calcium ions (Ca+2)
V	Extrinsic and Intrinsic	Proaccelerin, labile factor, or accelerator globulin
VII*	Extrinsic	Serum prothrombin conversion accelerator, stable factor, or proconvertin
VIII	Intrinsic	Antihemophilic factor, antihemophilic factor A, or antihemophilic globulin
IX	Intrinsic	Christmas factor, plasma thromboplastin component, or antihemophilic factor B
X	Extrinsic and Intrinsic	Stuart factor, Prower factor, or thrombokinase
XI	Intrinsic	Plasma thromboplastin antecedent (PTA) or antihemophilic factor C, Protein C
XII	Intrinsic	Hageman factor (HMK), glass factor, contact factor, or antihemophiliac factor D
XIII	Common	Fibrin-stabilizing factors (FSF)
* There is no factor VI. Prothrombinase is a combination of activated factors V and X.		

three basic mechanisms help prevent blood loss: (1) vascular spasm, (2) platelet plug formation, and (3) blood coagulation (clotting). These mechanisms are effective in preventing hemorrhage in smaller blood vessels, but extensive hemorrhage or wound infection requires medical treatment.

First, a **vascular spasm** occurs. When a blood vessel is damaged, the smooth muscle in its wall contracts immediately. Such a vascular spasm reduces blood loss for several minutes to several hours, during which time the other hemostatic

mechanisms go into operation. The spasm is caused by sympathetic reflexes involving pain receptors due to damage to the vessel wall. Signals are sent for the **extrinsic pathway** to begin.

When platelets come into contact with parts of a damaged blood vessel, their characteristics change drastically. They begin to enlarge and their shapes become even more irregular. They also become sticky and begin to adhere to collagen fibers in the wound. They produce substances that activate more platelets, causing them to stick to the original platelets. This accu-

The Clotting Cascading

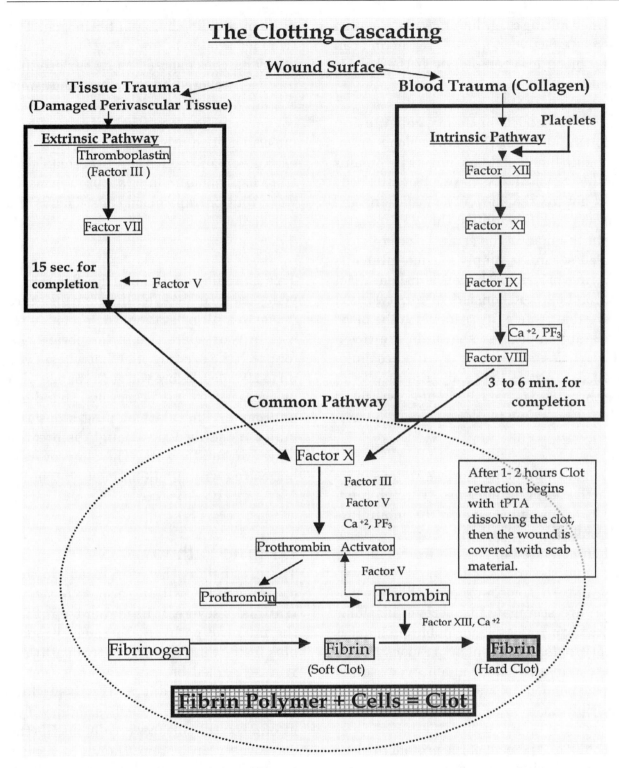

Figure 7.1. Outline of the blood clotting system. (Modified from Davis and Kenyon 1993 with permission)

mulation and attachment of large numbers of platelets to form a mass is called a **platelet plug**. The platelet plug is formed by platelets sticking to each other by collagen. Initially, it is soft, but it becomes hard when reinforced by fibrin threads formed during coagulation. A platelet plug is very effective in a small vessel and can stop blood loss completely if the hole is small.

After the platelet plug forms, **coagulation** (clotting) occurs. Normally, blood maintains its liquid state as long as it remains in the vessels. If it is drawn from the body and not treated, however, it thickens and forms a gel. Eventually, the gel separates from the liquid. The straw-colored liquid, called serum, is simply plasma minus its clotting proteins. The gel is called a clot (thrombus) and consists of a network of insoluble protein fibers in which the blood cells are trapped as shown on the cover. The process of clotting is called coagulation. If it occurs too easily, the result can be thrombosis—clotting in blood vessel. If the blood takes too long to clot, hemorrhage can result.

The body commonly stores enzymes in an inactive form for later use. Thrombin initially exists as inactive prothrombin. Prothrombin can cleave fibrinogen to fibrin when activated to become thrombin by a traumatic event to the skin or blood vessel. In turn, soluble proteins (like fibrinogen) are stimulated to become insoluble (like fibrin). It is the insoluble substances that produce the sealant responsible for the clot meshwork. In summary, dissolved or suspended substances become a solid precipitate.

About 2 to 3 percent of the protein in blood plasma consists of a protein complex called fibrinogen, a protein waiting to react. Normal fibrinogen is dissolved in plasma, like salt is dissolved in the Gulf of Mexico. Then another protein, called thrombin, slices several small pieces from two of the three pairs of protein chains in fibrinogen. The trimmed protein is now called fibrin. It has sticky patches exposed on its surface that has been covered by the pieces that were cut off. The sticky patches are precisely complementary to portions of other fibrin molecules. Long threads form, cross over each other, and (much as a fisherman's net traps fish) make an intricate protein meshwork that entraps blood cells. This is the initial clot.

This is analogous to having a flat tire in your car during travel. In order to secure travel to a filling station, you probably resort to your spare tire, a temporary one (or patch the existing one) that will get you up to 20 miles, before you get a change with permanent tire. (It is merely a band-aid approach.) The platelet plug stops the major bleeding long enough before the other can totally work. The plug covers a large area with a minimum of protein.

Extrinsic Pathway

The extrinsic pathway of blood clotting occurs rapidly, within seconds if **tissue trauma** is severe. It has fewer steps than the intrinsic pathway, therefore it operates faster than the alternative pathway. It is so named because the formation of prothrombinase is initiated by **tissue factor** (TF), also called thromboplastin, which is found on the surfaces of cells outside the cardiovascular system. Damaged tissues release tissue factor, (Figure 7.1) following several additional reactions that require calcium ions (Ca^{+2}) and several coagulation factors. Tissue factor is eventually converted into prothrombinase. This completes the extrinsic pathway and stage

1 of clotting. In stage 2, prothrombinase and Ca^{+2} convert prothrombin into thrombin. In stage 3, thrombin, in the presence of Ca^{+2}, converts fibrinogen that is soluble, to fibrin, which is insoluble.

Intrinsic Pathway

The **intrinsic pathway** of blood clotting is more complex than the extrinsic pathway and it operates more slowly, usually requiring several minutes because of its many steps. The intrinsic pathway is so named because the formation of prothrombinase starts on the surfaces of endothelial cells that line blood vessels, cells within the cardiovascular system. The intrinsic pathway is triggered when blood comes into contact with the damaged endothelial cells, or **blood trauma**. The damaged cells expose collagen and activate the lengthy pathway. The trauma damages the platelets, causing them to release phospholipids. Then, following several additional reactions that require the presence of Ca^{+2} and several coagulation factors, prothrombinase is formed. This completes the intrinsic pathway and stage 1 of clotting. From this point on, the reactions for stages 2 and 3 of the intrinsic pathway are similar to those of the extrinsic pathway. Once thrombin is formed, it causes more platelets to adhere to each other, resulting in the release of more platelet phospholipids. This is another example of a positive feedback cycle. Once the clot is formed, it plugs the ruptured area of the blood vessel and thus prevents bleeding. Permanent repair of the blood vessel can then take place. In time, fibroblasts form connective tissue in the ruptured area and new endothelial cells repair the lining of the blood vessel.

Clot Retraction and Fibrinolysis

Normal hemostasis involves two additional events after clot formation, clot retraction and fibrinolysis. Clot retraction is the tightening of the fibrin clot. The fibrin threads attached to the damaged surfaces of the blood vessel gradually contract because of platelets pulling on them. As the clot retracts, it pulls the edges of the damaged vessel closer together. Thus the risk of hemorrhage is further decreased.

The second event following clot formation is fibrinolysis, breaking up of the blood clot. When a clot forms, an enzyme called plasmin is formed. This enzyme can dissolve the clot by digesting fibrin threads and inactivating the fibrinogen, prothrombin, and coagulation factors. In addition to dissolving large clots in tissues, plasmin also removes very small clots in intact blood vessels before they can grow and impair blood flow to body cells and tissues.

Clot formation is a vital mechanism that prevents excessive loss of blood from the body. To form clots, the body needs calcium and **vitamin K**. Vitamin K is not involved in actual clot formation but is required for the synthesis of prothrombin and certain coagulation factors. This vitamin is normally produced by bacteria that live in the colon. Applying a thrombin or fibrin spray on a rough surface such as gauze may also promote clotting.

Hemostatic Control Mechanisms

Even though thrombin has a positive feedback effect on producing a blood clot, clot formation occurs locally at the site of damage; it does not extend beyond the wound site into the general circulation. One reason for this is that some of the coagulation factors are carried away by blood flow so that their concentrations are not high enough to bring about widespread clotting. In addition, fibrin itself has the ability to absorb and inactivate up to nearly 90 percent of thrombin formed from prothrombin. This helps stop the spread of thrombin into the blood and thus inhibits clotting except at the wound. This is an important design feature.

A number of substances that inhibit coagulation are present in blood. Such substances are called anticoagulants. Heparin, for example, is an anticoagulant produced by mast cells and basophils located in endothelial cells lining blood vessels. It inhibits the conversion, of prothrombin to thrombin, thereby preventing blood clot formation.

Clotting in Blood Vessels

Even though the body has anticoagulating mechanisms, blood clots sometimes form in vessels. Such clots may be initiated by roughened endothelial surfaces of a blood vessel as a result of atherosclerosis (accumulation of fatty substances on arterial walls), trauma, or infection. These conditions induce platelets to stick together. Clots in blood vessels may also form when blood flows too slowly, allowing coagulation factors in local areas to increase in concentration and initiate coagulation.

Clotting in an unbroken blood vessel (usually a vein) is called thrombosis. A thrombus may dissolve spontaneously, but if not, there is the possibility it will become dislodged and carried in the blood. If the clot occurs in an artery, the clot may block the circulation to a vital organ. A blood clot, bubble of air, fat from broken bones, or a piece of debris transported by the bloodstream is called an embolus. When an embolus becomes lodged in the lungs, the condition is called pulmonary embolism.

A Brief Summary of Blood Clotting

Blood clotting entails many aspects:

1. Hemostasis refers to the stoppage of bleeding.
2. It involves vascular spasm, platelet plug formation, and blood coagulation.
3. In vascular spasm, the smooth muscle of a blood vessel wall contracts to slow blood loss.
4. Platelet plug formation is the clumping of platelets to stop bleeding.
5. A clot is a network of insoluble protein fibers (fibrin) in which formed elements of blood are trapped.
6. The chemicals involved in clotting are known as coagulation (clotting) factors.
7. Blood clotting involves a series of reactions that may be divided into three stages: formation of prothrombinase (prothrombin activator), conversion of prothrombin into thrombin, and conversion of soluble fibrinogen into insoluble fibrin.
8. Stage 1 of clotting is initiated by the interplay of the extrinsic and intrinsic pathways of blood clotting.
9. Normal coagulation requires vitamin K and also involves clot retraction and fibrinolysis.
10. Anticoagulants (for example, heparin) prevent clotting.
11. Clotting in an unbroken blood vessel is called thrombosis.

One way to understand blood clotting is to compare it with a car engine. Most biochemical systems are composed of interrelated parts and components that must act simultaneously for the necessary functions in the system to work. These complex physiological systems in the body operate much like a car engine in which the spark plugs, pistons, radiator, fan belt, and other parts must operate together (Davis and Kenyon, 1993).

The formation, limitation, strengthening, and removal of the blood clot is an integrated biological mechanism. Problems with any single component can cause the entire mechanism to fail. This is analogous to the car engine, which fails to operate if the fan belt is missing, the distributor cap is cracked, or the spark plugs are clogged. The lack of certain blood clotting factors, or the production of defective factors, often results in serious health problems or death. The most common form of hemophilia occurs because one of these blood factors is missing. How could the blood clotting mechanism in humans evolve step by step over millions of years under the action of nothing more than random physiochemical forces?

Another way to understand the irreducible complexity of biochemical systems is to consider the interlocking components of a mousetrap (Behe, 1996). The mechanism of blood clotting is much like a mousetrap that has five parts: a catch, a platform, a holding bar, a hammer, and a spring. When assembled there is no gradual improvement of function. It does not work until every part is in place. The same thing is true inside a living cell and in each specific organ of the human body. Many body systems just will not work unless every part is present at the same time.

Furthermore, the success of blood clotting depends critically on the rate at which the different reactions occur. An organism would not long survive if the protein activated by thrombin formed at a significantly faster rate than proaccelerin (Figure 7.1). Blood clotting depends not only upon the occurrence of these reactions, but also on the correct timing and the rates of their actions.

Time Needed for Clotting

We have already determined the number of interdependent reactions for clotting to take place exceeds 30. The probability of 30 irreducibly complex reactions occurring simultaneously over evolutionary time approximates zero. If this were not enough to demonstrate the improbability of sequenced order of reactions and chance mutations to produce an intricate clotting mechanism, we introduce also the factor time length for each reaction set to take place. Not only must all the right chemicals be in place, but the timing of their interaction to produce the 'right' cascade is also vital for success (Figure 7.1).

A number of laboratory tests are used to evaluate the efficiency of coagulation. Normally, the bleeding of fingerstick should stop within to 2 to 3 minutes and sample of blood in a clean tube should clot within 15 minutes. Other techniques are available that can separately assess the effectiveness of the intrinsic and extrinsic mechanisms. The extrinsic pathway requires fewer steps to activate Factor X than the intrinsic mechanism does. It is a shortcut to coagulation, therefore, it takes about 15 seconds to take place. On the other hand it takes 3 to 6 minutes for a clot to form by the intrinsic pathway. For this reason, when a small wound bleeds you can stop the bleeding by massaging the site. This releases thromboplastin from the perivascular tissues and activates or speeds up the extrinsic pathway. The cascade of enzymatic reactions acts as an amplifying mechanism to ensure the rapid clotting of blood.

Eventually after the clot has formed, pseudopods of the platelets adhere to strands of fibrin threads and draw the edges of the broken vessel together, like a drawstring closing a purse. Through this process of clot retraction, the clot becomes more compact within 30 minutes. Finally within 2 to 6 hours later, we see the clot receding giving way to scar tissue and the wound is healing. In summary, the nature of blood clotting provides one of the strongest evidences of intelligent design and *Ex Nihilo Creation*. Only a Wise Creator could have ingeniously made a pathway so complicated and yet so effective in preserving life. For *life is in the blood.*

If these interrelated mechanisms began to originate in a slow, creeping fashion by microevolutionary changes, the result would be disastrous. Any member missing in the "molecular team" would have prevented the final mechanisms from working and most likely would have proved lethal before the end result was achieved by natural selection. It is more logical to believe that this irreducibly complex set of biochemical reactions was placed there by the actions of the Master Planner.

Remember the Creator formulated not only the "plan" for the blood cascade but also produced the first working organisms. He is not only the chief architect of the blood cells, proteins, and platelets, but also the Manufacturer of the components. He keeps everything going because He is the Maintainer. The predictable order of the coagulation cascade exists because the order of a precise plan was produced by intelligent cause. These finely-tuned and interdependent biochemical interactions are examples of what biochemist Behe calls irreducible complexity. Most creation scientists would go a step further and say this is clear, physical evidence of fingerprints from the Master's hand.

Chapter Eight
"All or None" Systems in the Human Body

The hearing ear and the seeing eye, the Lord hath made them both.

Proverbs 20:12

Like the multifaceted systems discussed in Chapter 6, the parts of the eye and the ear will function only as an entire unit or not at all. These interrelated parts might be called the "all or none" systems in terms of survival. The multiple parts of the eye and the ear must work fully in place and together in precision, or the function is impaired. In order for the eye or ear to perform optimally, coordination of all the parts is required.

The way these systems operate reminds us of the well accepted **all-or-none principle** in muscle and nerve physiology. In skeletal muscles, individual fibers contract to their fullest extent or not at all. A minimal threshold must be made in order for any action to take place. It either contracts, or not at all. In neurons, if a stimulus is strong enough to initiate an action potential, a nerve impulse is propagated along the entire neuron at a constant strength.

The Eye—A Living Camera

The eye's anatomy and optics operate like a camera and the eye's nervous system is comparable to a supercomputer. Yet, neither of these analogies is sufficient to fully describe the wonder of the human eye's structure and function.

The eye is more than any sum of machines that humans can construct!

The eye is built like the camera except that it is infinitely more sophisticated. Some modern cameras have auto focus and automatic adjustment of the iris diaphragm. The eye has a circular iris muscle that controls the opening called the pupil. In the eye, the lens can also change its shape to focus the light rays on the retina that is comparable to the film of the camera. The lens is made of living cells that are transparent. The cornea, the window-like membrane that covers the eye, is also transparent.

The most amazing component of the camera eye is its "film" which is the retina. This light sensitive layer at the back of the eyeball is thinner than a sheet of plastic wrap and is more sensitive to light than any man-made film. The best camera film can handle a ratio of 1000-to-1 photons in terms of light intensity (Menton, 1991). By comparison, human retinal cells can handle a ratio of 10 billion-to-1 photons over a dynamic range of light wavelengths of 380-to-750 nanometers. The human eye can sense as little as a single photon of light in the dark! In bright daylight, the retina can bleach out, turning its "volume control" way down so as not to overload. The light-sensitive cells of the retina are like an extremely complex high gain amplifier which is able to magnify sounds more than one million times.

There are over 10 million rods and cones in the retina and they are packed together with a density of 200,000 per square millimeter in the highly sensitive central fovea, the area of highest concentration of cone cells. These photoreceptor cells have a very high rate of metabolism and must completely replace themselves about every seven days! If you look at a very bright light such as the sun, they immediately burn out, but are rapidly replaced in most cases. Because the retina is thinner than the wavelength of visible light, it is totally transparent. Each of its minute photoreceptor cells is vastly more complex than the most sophisticated computer (Menton, 1991).

It has been estimated that 10 billion calculations occur every second in the retina before the image even gets to the brain! It is sobering to compare this performance with the best output of the most powerful computer. Stevens (1985, p. 102) reported that:

To simulate 10 milliseconds of the complete processing of even a single nerve cell from the retina would require the solution of about 500 simultaneous nonlinear differential equations one hundred times and would take at least several minutes of processing time on a Cray supercomputer.

Because there are about 10 million such cells interacting with each other in many complex ways, it would take something like one hundred years of computer time to simulate what takes place in the eye many times each second!

What makes this comparison even more striking is that nerve cells in the retina (rods, cones, bipolar neurons, and ganglion neurons) conduct their electrical signals approximately one million times slower than the conduction in circuit traces of "wires" in a man-made supercomputer. If it were possible to build a single silicon chip that could simulate the retina using currently available technology, it would have to weigh about 100 pounds while the retina weighs less than one gram. The "super chip" would occupy 10,000 cubic inches of space, whereas, the retina occupies 0.0003 cubic inches. The power consumption of the super chip would be about 300 watts, whereas the retina consumes only 0.0001 watts. It is amazing how the efficiency of the retina far exceeds the best computer chip that man can produce!

Darwin (1979, p. 217) once said that the very thought of the complexity of the eye gave him the chills. Attempts to explain the evolution of the eye, like most other evolutionary "explanations," are scientifically untestable. One must account not only for the eye, but also for an optically transparent membrane called the cornea, which bends the light rays before they enter the eye.

Together with the eye, the visual cortex of the brain, which is connected to the cerebrum, must translate optical information. The information begins as nothing more than differences in the amplitude and length of light rays and is then converted into what is perceived as three-dimensional color vision in real time. There is undoubtedly a scientific mechanism for all of this signal processing, and we now know much about it, but we are no closer to a scientific explanation of how eyes evolved in the first place than we were 150 years ago.

The human eye is an example of "compound traits" having parts that are interdependent upon each other and that do not function effectively apart from the other traits. It takes many muscles with the several eye parts (such as the cornea, the pupil, the lens, and the retina) working together, to get clear vision.

To suppose that the eye, (with so many parts all working together) ...could have been formed by natural selection, seems, I freely confess, absurd in the highest degree

Darwin (1979, p. 217) wrote the above just before he proceeded to express a rationale for how it might have happened anyway by small changes. But Darwin's argument was no explanation at all.

The human eye is among the most complex and sophisticated systems in the universe. Yet, in a critique of creation views called *Life's Grand Design*, Miller (1994) describes the human eye as a "flawed design" compared to the cephalopod eye. As radically different as invertebrates are from vertebrates in their other organs, the cephalopod octopus has an eye strikingly similar to that of man (Menton, 1991).

Several evolutionary biologists claim that the vertebrate eye is functionally suboptimal, because its photoreceptors in the retina are oriented away from incoming light (Dawkins, 1996a; Miller, 1994). They make these claims because the optic wiring and support vessels of the human eye are located in front of the photoreceptor cells in the retina. These biologists claim this arrangement degrades visual quality by scattering incoming light and creating a blind spot where the wiring must poke through the retina to reach the brain. The cephalopod eye has its retina in front of the "wiring."

The critics believe that evolutionary processes produced a superior eye plan in the octopus and squid. They reason that evolutionary tinkering through natural selection must have produced both the cephalopod and human eye; eyes that are optimal and suboptimal, respectively.

On the surface, their argument seems logical. Yet the human eye is much more sophisticated than the cephalopod eye. The physiology of human vision reveals great wisdom in this unexpected design, it provides for seeing objects in stereoscopic color. There are excellent functional reasons for human photoreceptors to be oriented in front.

Photoreceptor structure and function is maintained by a critical tissue, the retinal pigment epithelium (RPE) that is located beneath the retina (Figure 8.1). The RPE recycles photopigments, removes spent outer segments of the photoreceptors, provides an opaque layer to absorb excess light, and performs additional functions (Ayoub, 1996, p. 19).

The RPE is a fully functional tissue for the eye and must be located in back of the retina's photoreceptor cells in order for optimal vision. If all the wiring and support vessels were behind the retina, this would leave no room for the RPE. Optimality theory predicts minimizing cost and maximizing benefit. Although there is some extra energy cost to the "wiring" of the nerves in front of the retina,

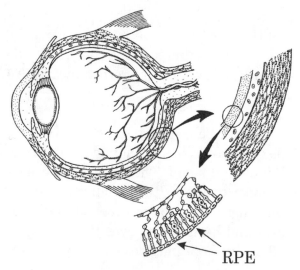

Figure 8.1. Retinal pigment epithelium (RPE).

the benefit of the RPE next to and behind the retina for maximum protection of RPE and for the superior photoreceptor functioning proves to be the best overall retina design for an eye.

The RPE must lie between the choroid and the bipolar cells (Figure. 8.1). So the human eye is not flawed as Miller contends. Rather, the conclusions of evolutionary biologists have demonstrated *flawed thinking* not *flawed design*! In summary, the RPE must be located outside the retina to minimize blood clotting and to supply the needs for the photoreceptors. If our eye like the cephalopod eye had no RPE, then our vision would be greatly reduced.

Would you want to trade your eyes for those of a squid or an octopus? Cephalopod eyes are extremely nearsighted, somewhat colorblind, and unlikely to form sharp images as our eyes can. The cephalopod has eyes designed for life in the deep oceans. The human eye, like the eyes of other vertebrates, has vision that is superior to cephalopods in the air. These finer

details of the structure and function in the human eye have been ignored by evolutionists when they say the human eye is suboptimal to the eye of a cephalopod.

The original design of the human eye is optimal. Its coordinated details make its structure more advanced than the Cephalopod eye, giving humans greater precision and visual acuity. The "imperfections," or eye impairments that some experience in vision (nearsightedness, farsightedness, astigmatism) are the result of the fallen nature of man, and are not part of the original plan. Upon closer inspection of the "reversed wiring" parts, we find there is "wisdom" in such a scheme.

The compound traits of the human eye make a fascinating system that requires that all its parts function together. This is another example of Paley's argument: Just as a watch must have had a watchmaker, there must have existed at some time and at some place a Divine Artificer of the human body.

The Hearing Ear

The human ear can detect sound frequencies of about 20 to 20,000 vibrations/sec. The ear also qualifies as an "all or none" system because it has many interdependent parts. Like vision, hearing takes place only after a cascade of mechanical, electrical, and biochemical events has occurred. There are numerous canals, fluids, and bones in the ear that transfer mechanical waves and convert them to meaningful sounds. One of the characteristics that humans share with other mammals is the presence of three small auditory ossicles (ear bones) and an outer, fleshy auricle.

The outer ear is made to receive audible signals and to transfer these vibrations to the eardrum and to the middle ear (see Figure 8.2). The three little bones are the malleus (hammer), incus (anvil), and stapes (stirrup). They are the smallest bones in the human body. They were named for their resemblance to tools used in the 19th century in the western United States. They transfer and amplify sound impulses through the middle ear. These ossicles are located within the cavities of the middle ear, which is within the temporal bone cavity. The three bones are attached to the wall of the tympanic cavity by tiny ligaments and are covered by mucous membrane. These bones bridge the eardrum with the inner ear, transmitting vibrations between these parts.

Although some people think of the outer ear (auricle, external auditory canal, pinna, etc.) as the site of hearing, it is the middle and inner ears that actually do the work. The middle ear has the tympanic membrane on its the outer side and the cochlea on its the inner side. The malleus is attached to the tympanic membrane and the vibrations are transmitted from it via the malleus and incus to the stapes. The stapes in turn is attached to the membrane in the cochlea that is called the oval window and it moves in response to vibrations of the tympanic membrane. When the stapes presses the oval window into the cochlea, another flexible membrane in the cochlea window bulges outward to relieve the pressure.

The fact that vibrations of the tympanic membrane are transferred through three bones instead of merely one bone affords protection. If a sound were too intense, the auditory ossicles might buckle. This protection is enhanced by the action of the stapedius muscle that attaches to the neck of the stapes. When the sound becomes too loud, the stapedius muscle dampens the movement of the stapes against the oval window. This action helps to prevent nerve damage within the cochlea when sounds repeatedly reach high amplitudes. In the

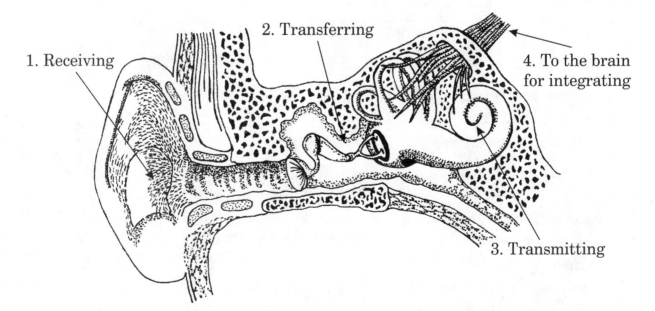

Figure 8.2. Functions of the ear.

case of a gunshot, however, the stapedius muscle may not respond quickly enough to prevent nerve damage.

The inner ear contains the most critical part of the hearing mechanism; the **organ of Corti**. It is situated in the snail shaped cochlea that is filled with **perilymph**. The cochlea has a twisting interior that is studded with thousands of hairlike nerve cells, each one of which is tuned to a particular vibration. When the stapes of the middle ear receives signals from the eardrum, such as a trumpet sound, the perilymph fluid inside the cochlea is set vibrating as a tidal current goes though it and transmits the signal to the auditory nerve and the brain.

Above the cochlea are three minute, fluid-filled semicircular canals. The **endolymph** in the semicircular canals provides us with balance but the canals do not play a role in hearing. If we start to fall, the fluid in one of the canals is displaced. Hair cells there detect a change and signal the brain that we are off balance.

How does the complex ear anatomy correlate its functions of receiving, transferring, transmitting, and integrating mechanical waves into nerve impulses that the brain perceives as sound (Figure 8.2)? Vibrating objects, such as the vocal cords of a person speaking, create percussion waves in the surrounding air. The waves from the air strike the tympanic membrane and are transferred to the middle ear through the vibrations of the eardrum. The three bones of the middle ear then amplify and transmit these movements to the inner ear. The inner ear converts the energy of pressure into mechanical waves and these waves are sent to the oval window. Vibrations of the oval window produce pressure waves in the fluid within the cochlea and in turn the auditory nerve is stimulated. The important mechanical and electrical steps involved in the cascade of events necessary for hearing are outlined in Table 8.1.

Table 8.1. Interdependent systems in the ear.

1. Sound waves enter the external auditory canal striking the tympanic membrane.

2. Waves pass through the ossicles and strike the oval window.

3. This sets up waves in the perilymph, where they strike the vestibular membrane and scala tympanic.

4. In turn, the increased pressure in the perilymph cause waves to strike the basilar membrane, and stimulate hairs in the organ of corti.

5. These hair cells convert a mechanical force into a receptor potential. The cochlea transduces the energy of the vibrating fluid into action potentials.

6. The hair cells release a neurotransmitter that initiates an impulse to the cochlear branch of the vestibulocochlear (VIII) nerve. This nerve terminates in the thalamus portion of the brain.

Movement of the perilymph within the organ of corti rotates the hairs to be displaced, thus stimulating nerve cells. These vibrations of the fluid over the hair cells are similar to the action of skilled fingers plucking the strings of a harp. The vibrations of the hair cells produce electrical signals that go to the auditory nerve. Within the organ of corti, sensitive hair cells detect minute pressure waves and convert them to electrical signals and this change from pressure to electrical waves is called **transduction.** In humans, each pitch of sound produces a maximum vibration in hair cells located at one point along the cochlea.

It is the brain that makes the interpretation and "sense" of the signal. The brain "knows" the pitch of the sound because it knows the location of the sensory neurons that are firing in response to signals from hair cells. The cochlea feeds thousands of these electrical messages from the ear to the brain, which then unscrambles the meaning of the sounds. This coordinated sequence of events is quite amazing.

The fact that all parts of the ear are necessary to produce hearing should be obvious when one considers the complex chain of mechanical and electrochemical processes involving the outer ear, the middle ear, the inner ear, the auditory nerve and the brain. Take away any of the bones, fluids, or mechanical hairs and hearing is impaired or deafness may result. Could such complexity arise by chance mutations and natural selection? We think not. The ear, along with the eye appears to carry the signature of an Intelligent Designer.

In the 19th Century, the famous comparative anatomist, George Cuvier, spoke of interrelated systems like these as having "a correlation of parts." Like Paley, he saw obvious signs of God's handiwork in the ear and in the human eye. Cuvier was a firm creationist and he advanced these arguments against those who believed in evolution (Morris, 1988). In the 20th Century, many creationists have argued for design in light of the interdependence of body structures. According to Behe (1996), as the number and quality of the related components in a system increases, we can be more and more confident that the systems arose by design. Our conviction grows that Solomon was correct in his writing (long before Paley, Cuvier, or Behe) that the eye and the ear testify to God's craftsmanship:

The hearing ear and the seeing eye, the Lord hath made them both.

Proverbs 20:12

Chapter Nine
The Wonder of Adaptation

*Natural selection operates essentially to enable the organisms to **maintain** their state of adaptation rather than to improve it.*

R. C. Lewontin (Parker, 1996, p. 84)

In the 20th century, the word **adaptation** has been associated with permanent structural changes in plants and animals changing over time in an evolutionary progression. The word adaptation, however, may also be used to describe short-term physiological changes that take place in an individual's creature's life to monitor and adjust to changing environmental conditions. This second definition of adaptation is one that creationists use and it fits better with the data observed in nature.

The evolutionary perspective sees adaptation as a continuum from the acclimation previously described to major adaptive radiation of primates into modern day apes and humans. Evolutionists see adaptation as something new on the scene, selective advantages made in a struggle for survival. They see progress in adaption through natural selection, long time periods, chance, and blind tinkering by mechanistic forces. The evolutionist sees man's lowly origin in an apelike ancestor, such as *Ramapithecus*. Through various adaptations, these apelike creatures gradually stood upright and evolved into early man such as *Homo habilis* or *Homo neanderthalensis* and then into *Homo sapiens* (Figure 9.1).

Adaptation Means Adjustment

It is amazing what the body is able to en-

Figure 9.1. Hypothetical monkey-to-man concept.

dure. Imagine if we could not adapt to sudden changes in temperature, or altitude, or if we could not tell the constant changes in friction on the palms of our hands and soles of our feet. The Creator has designed our body to adapt to a wide variety of difficult, new conditions. Subtle responses constantly occur in our body automatically such as the dilation of the pupils in response to bright light or the increase in the heart's output when we suddenly stand up. Sometimes we lose our balance because the heart does not increase cardiac output quite fast enough. These features are called **short-term physiological adaptations**. There is no indication that these adaptations resulted from the natural selection of random mutations.

Short-Term Physiological Adaptations

Try to imagine what life would be like if we had to make numerous subtle, physiological adjustments voluntarily on a constant, minute-by-minute basis. We would not have much time for anything else. Furthermore, if we increased our blood pressure, we could have kidney damage or possibly a stroke. People who stood suddenly would faint because of blood pressure changes in response to gravity. If we failed to decrease our pupil circumference in bright light, we would permanently damage the retina. If callouses did not form on our skin, it could mean constant blisters and raw, open wounds subject to infection.

Physiologists understand well just how detailed these subtle adjustments must be. The Creator has given people (and all other vertebrates) an autonomic nervous system (ANS) composed of a sympathetic and parasympathetic division. These two cooperate in opposite functions with each other to ensure a high degree of adjustment. Our skin looks "flushed" or red on a hot day because our ANS causes a vast network of tiny blood vessels in the dermis of the skin to dilate (increase in circumference). This is the Creator's way of ridding the body of excess heat from subdermal tissues. Such a process, along with body perspiration, causes us to maintain a constant temperature. We take these cooling responses for granted, but if they stopped, dangerous overheating and possibly death would follow.

Consequences of Deleterious Mutations

Bodily functions and adjustments frequently work in a cascade effect. Can mutations (random genetic mistakes) and a trial and error selective process be the source and final explanation for such precise homeostatic regulating systems? What is known about mutations shows they are disruptive, random aberrations.

Think of what would happen if someone experienced the mutation of a key enzyme involved with the breaking down of glycogen into glucose in the liver. Any such mutation in these examples would spell death or a serious problem for the individual. Mutations are basically deleterious.

All parts of any system in an individual must work together harmoniously from the beginning to survive. Energy demands must be constantly monitored by the liver. The liver makes precise biochemical adjustments, releasing just the right amounts of blood glucose into our circulation. It does this by breaking down glycogen (animal starch), a storage form of glucose. If we do not eat, then the body must find other forms of energy before the glycogen reserve is depleted. The body is designed to draw from our fat reserves before our glycogen reserve is exhausted. The Creator has programmed the body in ways such as these, that we are just beginning to understand.

Simply put, life would not be possible if we had to consciously monitor all of our internal activities. All our time would be spent consciously making subtle adjustments throughout our bodies so we would not die. People would have to think about physiological details in order to survive and would not be able to lead a "normal" life as we know it today. But according to evolutionary theories, these important adjustments were developed by trial and error. Making one major error, however, would cause an individual to die.

It is clearly more reasonable to assume that there were designed controls built into each individual's genetic makeup (genome). As the individual grows and develops in the womb, the "controls"(ANS) are programmed to develop, helping the growing child to maintain its safe biochemical balance from the very start!

Adaptation to Higher Altitude

Another adaptation that frequently affects world class athletes is travel to various arenas in high altitude cities. This example is discussed in depth to help you appreciate the wisdom of the Creator and how He cares about the details of your life.

As one goes higher in altitude, air molecules become less concentrated and more spaced apart. This means we must breathe harder to bring in the same amount of oxygen that we take for granted at sea level. The production of hemoglobin, a red blood protein designed to carry oxygen, must be increased; and therefore, an increase of red blood cell production is also needed.

This increased production is in response to rarefied air. But how does our body "know" to do this? It is by a process that one would never guess. Our kidneys are sensitive to a decrease in oxygen levels of the blood. When this occurs, a hormone called erythropoietin is secreted from the liver, as well as the kidneys (Guyton, 1991; Hole, 1995). Erythropoietin stimulates bone marrow to produce more red blood cells which yield more hemoglobin and more binding sites for oxygen.

At high altitudes, where the pressure of oxygen is reduced, the amount of oxygen delivered to the tissues decreases. In response to prolonged oxygen deficiency over a period of weeks, erythropoietin is gradually released, primarily from the kidneys and to a lesser extent from the liver. Erythropoietin travels via the blood to the red bone marrow and stimulates increased red cell production. The body's adaptation to high altitudes, including the formation of new red blood cells, takes place over many days.

Eventually the number of cells reaches a level that is sufficient to supply oxygen to these body tissues. When this happens, erythropoietin release ceases and red cell production is reduced (Guyton, 1991; Hole, 1995). This is one of the many wonders of adaptation!

Adaptation and Evolutionary Assumptions

An evolutionary theme is that the present adaptation, distribution, and abundance of organisms are products of both long-term evolutionary changes and ongoing interactions with the environment. According to the naturalistic view, various anatomical structures and physiological mechanisms have evolved as adaptions through random mutation and natural selection in response to changing environmental conditions. Organisms survive and reproduce in areas where environmental conditions to which they are adapted are found.

Sickle Cell Anemia: A Beneficial Mutation?

Certainly mutations can cause changes in structure, but the student should ask what kind of changes are produced. Are these changes good or bad? We rightly call mutations mistakes in DNA. It is unlikely that these random mistakes in the DNA produce good changes that would last for generations. About 99.9% of mutations are harmful, and many of these are fatal. One of the few examples that evolutionists frequently give of a beneficial mutation (positive change) in humans is the sickle cell trait.

Sickle cell anemia is a blood disease (carried by a recessive gene) in 8–11% of the black population in the United States. It is caused by an abnormal form of hemoglobin called hemoglobin S (Van de Graff and Fox, 1995). This defect is the result of an amino acid substitution (valine for glutamate) which is caused by a single DNA base change that codes for a different beta chain. Those diagnosed with sickle cell anemia carry low oxygen and struggle for existence and a normal life. They cannot participate in athletic activities (or even exercise) for long. If you think that the sickle cell trait is a beneficial mutation, interview anyone who has the disease.

The author Gillen worked in a predominantly Afro-American high school for four years and saw the emotional and physical pain that accompanies sickle cell anemia patients and their families. They do not look at this genetic mutation as an evolutionary advantage. These people are looking for an effective treatment, or cure.

The sickle cell mutation obviously carries a disadvantage in America. In Africa, however, carriers for sickle cell anemia have both hemoglobin A and hemoglobin S in their blood cells. They are more resistant to malaria than noncarriers. This is because the parasite that causes malaria, *Plasmodium vivax* cannot live in red blood cells that contain hemoglobin S. Therefore, carriers do have a better chance of not contracting malaria in Africa. Wherever malaria is rampant, there is an advantage to carrying the sickle cell trait.

Under low oxygen tension, HbS is less soluble than normal hemoglobin. Resistance to malaria appears to be the result of the sickling process. Upon infection by the mosquito-borne protozoan *Plasmodium falciparum*, erythrocyte pH is lowered by anywhere from 2% to 40% sickling in erythrocytes. Sickling changes the red blood cell's membrane permeability to potassium ions. Normal red blood cells maintain a potassium ion internal concentration higher than that of the surrounding blood. Upon sickling, the red blood cells lose potassium. At this stage in the parasite development, a high potassium concentration is required. The loss of the ion upon sickling deprives the parasite of the ion and the parasite dies (Paolella, 1998).

This variation in red blood cells for carriers is an example of a good adaptation under specialized conditions. But the consequences of sickle cell anemia are severe in individuals who are homozygous for this trait. In most circumstances, it is detrimental and frequently results in death. In most parts of the world, the sickle cell trait is being selected against. Only in malaria-infested areas does the sickle cell trait confer an advantage to survival.

What can the creationist conclude? Natural selection can (and sometimes does) slow the rate of genetic decay produced by random mutations but this hardly demonstrates that mutation/ selection produces upward and onward progress. We can conclude, however, that both this "built-in" genetic variability and short-term physiological adaptations are helpful to human survival in a changing world. They can be seen as examples of God's providence (provision).

Physiological Responses: Homeostasis and Acclimation

Physiological responses to environmental change are generally slower than behavioral responses although some may occur very rapidly. An example of a faster physiological change would be when blood vessels in the skin constrict within seconds to reduce loss of body heat when the skin is exposed to very cold air. Physiological adaptation is centered around regulation and homeostasis. Humans function most

efficiently under certain environmental conditions which are optimal for each individual. Efficiency declines both above and below optimal values.

Physiological responses to changing environments can shift tolerance limits of organisms. Acclimation is a gradual process and is related to the range of environmental conditions. Some organisms can react to environmental change with responses that alter body form or internal anatomy. Such forms of acclimation are usually reversible.

For example, changes in skin color are common in animals, where these oscillating color variations allow them to blend into their environment. Camouflage is a natural protection strategy seen throughout the animal kingdom, but most readily observed in reptiles. The familiar green anole (*Anolis carolinensis*) and the chameleon change color depending upon their background: green skin on greenish leaves and brown on tree trunks. However, genuine morphological change caused by adaptation is extremely rare (if existent at all) for humans.

Adaptation Over Time: Evolutionary Progress or Designed Acclimation?

It is important to remember that behavioral and physiological adaptations that can change over time are subject to natural selection. The question to ask is whether short-term adjustments actually accumulate into evolutionary vertical changes or do they simply reflect pre-planned designed acclimations.

According to evolutionists, natural selection also places constraints on the distribution of populations by restricting them to localized environments. Organisms adapted to one type of environment may not survive if dispersed to a foreign environment or may become extinct if the local environment changes beyond their tolerance limits. The existence of a species in a particular location depends on the species reaching that location and being able to survive and reproduce after getting there.

Design Perspective of Adaptation

In contrast, those who hold the design view see adaptation as preplanned, built in variability for animals or humans enabling them to survive in a changing environment. Creationists predict that most adaptations are reversible and short-term, whereas, evolutionists would predict that most adaptations are irreversible and long-term. To the creationist, adaptation includes the care and concern of the Creator for each individual (Parker, 1996). Designed adaptation includes purpose, plan, symmetry and interdependence. Real adaption must be planned ahead of time. The Creator helps His creatures to fit into various environments, meeting changes within the same environment by a flexibility built into the genetic code.

Changes that occur within kinds are examples of horizontal evolution and ones that can be readily observed. There are, however, no examples observed of changes among kinds that demonstrate increasing complexity. Vertical evolution would require gene expansion and major constructive macromutations. There is no evidence to date that such gene mutations have taken place. Creationists see adaptation as an acclimation to new environments within boundaries. They find patterns of regularity and order. In summary, creationists see adaptation as a provision for the creature that the Creator had already planned and foreseen from the beginning of time. The Bible summarizes it this way:

For we are his workmanship, created in Christ Jesus unto good works, which God hath before ordained that we should walk in them.

Ephesians 2:10

and

Know ye that the Lord He is God: it is He that hath made us, and not we ourselves.

Psalms 100:3

PhysioFocus 9.1
Blue Angels, Adaptation, and the Physiology of Flight

Have you ever wondered what it might be like to fly in a jet that can soar faster than the speed of sound? What happens to the body as you turn upside down, move suddenly sideways in Delta rolls, or soar at high altitudes as Blue Angel aviators do? The answer may surprise you. Blue Angel pilots train for over 3000 hours before they even begin their first show. Part of the reason is that they would "black out" if they did not acclimatize to the high "G" forces. During shows pilots frequently travel at speeds that exceed 500 mph. And in war, these jets can exceed 1190 mph in actual combat maneuvers. These daring pilots perform these feats only after extensive training. Their bodies must undergo many short-term adaptations before safely attempting such stunts.

When a jet makes a turn, the force of centrifugal acceleration is determined by the following relationship: $F = mv^2/r$; where F is the centrifugal acceleratory force, m is the mass of the object, v is the velocity of travel, and r is the radius of curvature of the turn. From this formula it is noted that as the velocity increases, the force of centrifugal acceleration increases in proportion to the square of the velocity. In addition, the force of acceleration is directly proportional to the sharpness of the turn (or inversely proportional to the radius).

Effects of Acceleratory Force

When a pilot is simply sitting in his seat, the force with which he is pressing against the seat, results from the pull of gravity, and it is equal to his weight. The intensity of this force is said to be 1 G because it is equal to the pull of gravity. If the force with which he presses against the seat becomes 5 times his normal weight during pull-out from a dive, the force acting upon the seat is 5 G. If the jet goes through an outside loop so that the pilot is held down by his seat belt, negative G is applied to his body, and, if the force with which he is thrown against his belt is equal to the weight of his body, the negative force is -1 G.

Effects of Centrifugal Acceleratory Force on the Body

The most important effect of centrifugal acceleration is on the circulatory system because blood is mobile and can be translocated by centrifugal forces. Centrifugal forces also tend to displace the tissues, but, because of their more solid structure, they ordinarily do not sag enough to cause abnormal function. When the Blue Angel aviator is subjected to a positive G, the blood is centrifuged toward the lower part of the body. Thus, if the centrifugal acceleratory force is 5 G and the person is in an immobilized standing position, the hydrostatic pressure in the veins of the feet is 5 times normal, or approximately 450 mm Hg. This is true even in the sitting position when the pressure is nearly 300 mm Hg. As the pressure in the vessels of the lower part of the body increases, the vessels passively dilate, and a major proportion of the blood from the upper part of the body is translocated into these lower vessels. Because the heart cannot pump unless blood returns to it, the greater the quantity of blood "pooled" in the lower body the less becomes the cardiac output. Figure 9.2 illustrates the effect of different degrees of acceleration in systemic arterial pressure for a Blue Angel aviator in the sitting position. It shows that when the acceleration rises to 4 G the systemic arterial pressure at the level of the heart falls to approximately 40 mm.

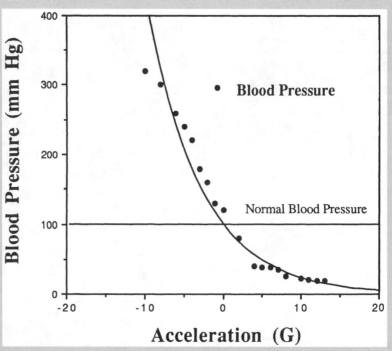

Figure 9.2. Effects of acceleration on aviator's blood pressure.

During rigorous flights, the three body systems primarily affected are the nervous, respiratory and circulatory systems. Most of these effects are the result of a reduced arterial pressure that causes reduced oxygen to the tissues (hypoxia); this in turn causes changes in nervous and respiratory systems. First, there is a proliferation of effects on the nervous system including, delay in reaction time, breathlessness, dizziness, dullness, drowsiness, euphoria, fatigue, headache, poor judgment, lightheadedness, faulty memory, muscle incoordination, numbness, performance deterioration, tingling, blurred vision, and vertigo. Then, a lack of nervous coordination aggravates more respiratory symptoms including hyperventilation, pale clammy appearance from hyperventilation, cyanosis (blue face) from hypoxia, middle ear may exhibit barotitis media, vertigo, sinuses blocked (barosinusitis), and possibly unconsciousness or death. These symptoms are all interrelated, such that one problem aggravates

the others in a cycle (Guyton, 1991).

The body can overcome many of these negative side effects through physical training. Over a period of six months to a year of rigorous flight training, aerobic exercise, good nutrition, and subjecting the body to varying G forces, the body adjusts. Apart from rigorous flight school, the body would most likely collapse or "black out" during a Blue Angel flight. Next time you see the Blue Angels fly in synchrony and make it look easy, just remember it was easy after several years of training and the body undergoing numerous physiological acclimatizations. This is not an evolutionary adaptation, but a built-in, preplanned variation in the body's marvelous adjustment to new situations. This acclimatization occurred by design, not by chance.

Chapter Ten
Understanding Man's Uniqueness

What is man that thou art mindful of him? ...For thou hast made him a little lower than the angels, and has crowned him with glory and honour.

David (Psalms 8:4–5)

The evolutionist sees modern man as a product of successive adaptations from a wild, forest environment to a cultured prairie. Evolutionists question the uniqueness of humans and see them merely as complex animals with which natural selection has tinkered over the eons. Is man unique, or is he just a "naked ape"? Man is classified by many evolutionists as a "naked ape" because he shares several physical similarities with the great apes and yet lacks hair over most of his body. Did we descend from the trees and become carnivores? If our ancestors began racing after game on the hot African savanna plains instead of staying in the shady tropical forests, this would have caused them to overheat, says the evolutionist. But if this is the explanation for our ancestors' loss of hair, why did lions and jackals not loose their hair, too? Nakedness is not essential for the hunter or the hunted.

Other macroevolutionists hold that the discovery of fire and clothing made hair expendable. But why would humans lose their fur, put on clothes, and sit near a fire if they were originally warmed by fur? This evolutionary theory of monkey to man backfires. There are major differences between humans and primates which (include monkeys, chimpanzees, gorillas, gib-bons, and orangutans). In this chapter, we will examine the singular and distinctive anatomy of human beings.

Biological Distinctions of Man

Only the human body stands and moves fully erect, having fully dexterous hands. People are alone among the land creatures in that they are naked (not covered with hair). We are not naturally protected against extremely cold temperatures, like the great apes. Instead we are designed for intimate contact with other humans. Only human beings are able to mate face to face, person to person. The human brain alone has a center for understanding and producing speech. We go though a longer childhood and a longer life than the other animals. These distinctives and other unique human traits should make us pause and carefully attribute our origin to the Creator.

Although our DNA may be more than 98.5% the same as chimpanzees, we are still very different. The 1.5% DNA difference between humans and chimps is what makes us genetically distinctive and different from these primates. Chimps lack a speech center and a long attention span. Although chimps build and use tools, they are limited in what they can construct.

They can accomplish physical tasks, like climbing trees and braving harsh weather conditions, but chimps cannot thread a needle or march in a parade playing a trumpet while standing fully erect. They have physical limitations, such as less dexterous hands. Their skeleton will not allow them to stand upright for a very long time.

Although humans are part of nature, our body is distinct from other creatures and it is the pinnacle of the Creator's design. Charles Darwin was convinced that man's body bore the stamp of a lowly origin. Creationists think that man is of a high origin. There are many differences between man and apes. The fossil record also indicates that these differences have always existed (Gish, 1995). Our bodies reflect a distinctive person, value, plan, purpose, design, and meaning. The structural differences in the human spine, pelvis, foot, thermoregulatory mechanisms, and the cerebral speech center are just a few examples of the extraordinary complexities

that appear fully developed only in humans. A few anatomical and physiological distinctions are given for man in Table 10.1.

In Jane Goodall's book, *In the Shadow of Man*, there is information that shows you how impressive the chimps are when compared with other animals. They use simple tools and can be taught to use simple language symbols. They go through periods of mating and mothering for six years. When chimpanzees are compared to monkeys and other creatures, they perform impressive feats. Yet, in study, the lives of chimps appear repetitious, dull, and monotonous by human standards. A few behavioral distinctions are given for man in Table 10.2. The human body tells of our special dominion, our personhood, and our God-given glory. Man has moral and ethical concerns, as well as an historical awareness found in no other creature. Instead of living out of the "shadow of the ape" as evolutionists assert, we see our bodies in a new light. It is special and glo-

Table 10.1. Anatomical and physiological distinctives of man.

1. Genetically 1.5% different from chimpanzees
2. Stands and moves fully erect with spine, pelvis and foot unique
3. Manual dexterity of hands having 70 muscles
4. Completely opposable thumbs
5. Relatively hairless: designed for contact, not for physical protection
6. Brain center for logical understanding and speech in the cerebrum
7. Unique throat for word use; well developed vocal cords
8. Most advanced thermoregulatory mechanism
9. Sex not associated with estrus
10. Stereoscopic vision: allows depth perception and three-dimensional viewing

Table 10.2. Behavioral distinctions of man.

1. Abstract thought

2. Languages of man

3. Technology and extensive toolmaking

4. Culture making

5. Reflective awareness

6. Ethical concern

7. Esthetic urges

8. Historical awareness

9. Metaphysical concern

10. Complex creativity (in the Designer's image)

11. Long, slow development and parental care.

rious. We hold that people did not evolve from sophisticated apes, but were created distinctly by an intelligent Creator.

Understanding Some Unexpected Features

One of many attacks on the planned creation theory is the alleged presence of supposedly "imperfect" and "vestigial" structures found in humans and other creatures. In preceding chapters, we have already discussed a few, such as the retina of the human eye, (perhaps the most complex tissue in the body) which evolutionists claim to be "wired backward."

According to the naturalistic reasoning of Drs. Steven Gould (1980) and Ken Miller (1994), this is the strongest case for the "tinkering" of evolution. According to these

biologists, if there is a Designer, then his methods are not very intelligent because he engineered many imperfect structures and flaws in nature. According to evolutionary thinking, these flawed designs and vestigial structures are best explained by chance mutations and blind physio-chemical forces. Natural selection is seen as the *blind watchmaker* with anatomical structures of living things over long periods of time.

One popular American biologist, Stephen Jay Gould (1980), states that the bony wrist protuberance that the panda uses to strip succulent bamboo shoots open, is a badly flawed design feature. If this is true, one would be tempted to kindly invite Gould to explain how an evolutionary development would create a better structure for the panda. Secondly, if it is so flawed, why have pandas not become extinct? Would not the creature have become extinct considering the importance of such a function? At least we ask if this is such a bad thumb, why has it not changed or been modified by evolution? Could it have been that it is not flawed at all and is well-suited for the purpose of opening bamboo?

In the botanical world there is a biochemical process called photorespiration. It has been maintained by many plant biologists that this process is very wasteful. They argue that if scientists could somehow circumvent photorespiration, more food could be produced by the plant. But recently two Japanese scientists have discovered a most important function for photorespiration: It protects the plant leaves from a devastating process called photooxidation. The dashboard of a car, when left in strong sunlight over the years, cracks and peels as a result of photooxidation. Photorespiration is involved with several other beneficial functions, such as the production of two amino acids (protein building blocks). So photorespiration is not wasteful (Campbell, 1996).

Pain Network: Flawed Design or Flawed Thinking?

Even for those who believe in a Creator, it has often been thought that one of God's mistakes was pain. Either the Creator is evil in allowing such physical pain to afflict humans, or his pain network in the nervous system was one big flaw. Superficially, there seems to be a case for such an argument. After all, who among us enjoys chronic back pain or likes watching a friend suffer with migraine headaches.

The answer to this observation will be found not by those who superficially glance at everyday life in America. But rather the answer is found among those who have observed people without the ability to sense pain, like those who suffer from Hansen's disease (formerly known as leprosy). Advanced leprosy patients are known for their "clubbed" hands and feet. These people have lost their sense of pain and have repeatedly subjected their limbs to severe pressures that led to permanent injury to their hands and feet all because they did not feel pain (Brand and Yancey, 1984; 1993).

Bodily adjustments can also work in a cascade effect such that one effect or adjustment must occur in order for the next adjustment to happen. If we did not receive the pain stimuli to blink, after several hours of not washing and lubricating the surface of our eyeball, it would dry out. Several days later our corneas would be damaged and blindness could occur, just because we did not do something as simple as blinking. This is sometimes the case of those who suffer from Hansen's disease. The microorganism, *Mycobacterium leprae*, destroys the pain receptors in the cornea, thus destroying the mechanism to blink (Brand and Yancey, 1980; 1984; 1993).

Clearly, pain is used by the Creator to protect our body from harming and ultimately destroying itself. Dr. Paul Brand, winner of the prestigious Al Lasker award for his humanitarian service to lepers, calls pain one of God's great gifts. He found that leprosy patients who lost their ability to sense pain would frequently injure themselves without knowing it. Have you ever said, "That hurts my eyes?" We know when to shield our eyes or look away from bright sources of light because of the photoreceptors in our eyes.

Foods that have a high temperature could severely damage the delicate lining of our mouth and throat. Heat can cause the throat to swell shut in extreme cases, possibly leading to death by suffocation. Think of that the next time you burn your lips on a hot drink! In summary, the pain network is an unexpected design created for our protection.

But once again, as science has further unlocked the mysteries of the biological world, functions have been found for the pain network in humans, the thumb in pandas, and the processes of photorespiration in plants. Each of these has a specific and unusual role to play.

Each of the so-called "imperfections" seems to be a product of unexpected designs, adaptations, and advantages. Are these unusual features actually part of flawed and imperfect structures as certain evolutionists believe?

Finally, the tone of a recent National Academy of Sciences (NAS) book (1998) indicates that the there is a succession of well-documented intermediate forms or species leading from early primates to humans. The tone of NAS is dogmatic, rather than tentative. It ignores the lessons of Piltdown man, *Ramapithecus,* other prehistoric men, and the uniqueness of the man revealed in Scripture and the points made earlier in this chapter. The NAS ignores problems in anthropology. In science, tentative conclusion should be stated in tentative form. Science has a bearing on our understanding of what it means to be human. Yet, science is not the only discipline that studies humans nor the only one with an important stake in that question. In all areas of science and especially when human origins are under consideration, we concur with an American Scientific Affiliation (ASA) statement:

The confidence expressed in any scientific conclusion should be directly proportional to the quantity and quality of evidence for that conclusion.

ASA, 1986, p. 42

In the matter of human origins, there remain many science mysteries. The National Academy of Sciences would have done well to have consulted with one of American's founding fathers, Benjamin Franklin, regarding the origin of man. Franklin, a nationally-recognized scientist during the earliest days of the United States, said:

I never doubted, for instance, the existence of the Deity that He made the world, and governed it by His providence.

Eidsmoe, 1987, p. 195

PhysioFocus 10.1
New Thinking on Split Brains

About 20 years ago, Gazzaniga discovered that each brain hemisphere controls vastly different aspects of thought and action (Gazzaniga, 1998). Each half has its own specialization, thus, has its own limitation and advantages. The left brain is dominant for language and speech. The right excels at visual motor tasks. The language of these findings has become part of our culture; writers are referred to as left-brained, and visual artists as right brained. Split-brain studies have continued and shed light on language, brain organization, perception, and attention. The original split brain studies suggested that the distinct halves could still communicate, but each had its own specialized action. The bridges of neurons that connect the hemispheres are called **commissures**. The **corpus callosum** is the most massive of these and typically the only one severed during surgery for epilepsy.

Studies of communication between the hemispheres have led to an important finding about the limits of nonhuman studies, in addition to helping neuroscientists determine which systems still work when the corpus callosum is severed. Typically neurobiologists have used animal models when trying to understand the human brain. For many years, neuroscientists have examined the brains of monkeys and other primates to explore the ways in which the human brain operates. It has been commonly believed among secular biologists that human brains have parallel "wiring" to apes. Charles Darwin emphatically disseminated the idea that the brains of our closest relatives have an organization and function largely similar, if not identical, to our own.

Recent split-brain research denies Darwin's assertion. The prediction of evolutionists that human and various ape brains are similar in construction has been demonstrated to be wrong. Humans are distinctly different in terms of brain hemisphere function. Although some structures and functions are remarkably similar among humans and other primates, differences abound. The anterior commissure provides a dramatic example of human distinction.

This small structure lies below the corpus callosum. When this commissure is left intact in otherwise split-brain monkeys, the animals retain the ability to transfer information from one hemisphere to the other. People, however, do not transfer visual information in any way.

Hence, the same structure carries out different functions in different species. The model of descent with modification for brain structure and function has been demonstrated to be wrong. On the other hand, the creation model for brain organization predicts anatomical and physiological distinctions among different kinds of animals. Another supporting case for the human uniqueness reported is one where a left-handed patient spoke out of her left brain after split-brain surgery (not a surprising finding in itself). But the patient could write only out of her right, nonspeaking hemisphere. This dissociation confirms the idea that the capacity to write need not be associated with the capacity for phonological representation. Put differently, writing appears to be an independent system, an attribute of the human species. Humans stand alone in being able to utilize language.

Chapter Eleven

Maintaining Boundaries

"Thou hast set a bound (boundary) that they may not pass over"

David (Psalms 104:9a)

We find in the Bible that the Creator has placed boundaries upon both physical and geologic phenomena (Job 38:10, 11; Jer. 5:22). He has placed genetic limits on variation within the various kinds of organisms (Genesis 1). The Creator has also placed limits in the human body for the protection and good of all creation. The boundary concept may be seen in limiting the entry and exit of chemicals and pathogens. Every organism must be able to maintain its boundaries so that its insides remain distinct from everything outside. In this chapter, the patterns involved in **maintaining boundaries**, will be illustrated with examples from the human body. First, we explore the broader concept of boundaries and then we describe specific boundaries that serve to keep pathogens out of the body.

Every human body cell is surrounded by an external membrane that encloses its contents, selectively allowing helpful inflow and harmful outflow of chemical compounds. In addition, boundaries protect the body's insides from outside pathogens. Each cell has a protective covering—a membrane that limits it. Of the approximately 75 trillion body cells, most are surrounded by friendly human cells that share the same adaptive functions and live in harmony. All these cells have barriers provided by the Creator to control what goes in and what comes out.

Anatomical Boundaries in the Body

In the human body, we can observe boundaries that limit the entry and outflow of chemicals, such as the membrane around the brain. Cells surrounding the brain are designed to regulate tight control over what enters the brain. The astrocyte cells primarily regulate substances that pass in or out of the brain. Astrocytes and other blood-brain-barrier cells ensure that the brain will not be subject to chemical fluctuations in the blood. Thus, they provide protection against dangerous chemicals. They work like a sieve, screening larger molecules from smaller ones at the brain's edge.

The intact skin and the mucous membranes that extend into the body cavities are among the most important resistant factors. Unless penetration of these barriers occurs, disease is rare. Penetration of the skin is a fact of everyday life. For example, a cut or abrasion allows staphylococci to enter the blood, and a mosquito bite acts as a hypodermic needle permitting many different organisms to enter of which yellow fever viruses, malaria parasites (*Plasmodium* sp.), certain rickettsiae, and plague bacilli are but a few examples. Other means of penetrating the skin include splinters, tooth extractions, burns, shaving nicks, war wounds, and injections.

Certain features of the mucous membranes provide resistance to parasites. Cells of the mucous membranes along the lining of the respiratory passageways secrete mucus that traps heavy particles and microbes. The cilia of mucous membrane cells move particles along the membranes and up the throat, where they are swallowed. This mechanism is referred to as the **mucociliary escalator.**

Stomach acid then destroys any swallowed microorganisms. A natural protection to the gastrointestinal tract is provided by stomach acid, which has a pH of approximately 2.0. Most organisms are destroyed in this environment. Notable exceptions include typhoid and tubercle bacilli, *Helicobacter pylori* (which is the ulcer bacterium), protozoan cysts, and also the viruses of polio and hepatitis A. Bile from the gall bladder enters the system at the duodenum and serves as a chemical barrier. Duodenal enzymes also digest the proteins, carbohydrates, fats, and other large molecules of microbes.

All parts of the body surface exposed to the outside world are covered with epithelial cells. These cells are packed tightly together and rest on a thin layer of noncellular material, the basement membrane. The parts of the body that are exposed to the outside world include not only the skin that is in direct contact but also the mucous membranes of the genitourinary tract, the respiratory tract, and the alimentary tract. Although they are inside the body they are nevertheless exposed to the external environment through the intake of food and air.

Each region of the body has, in addition to the tight packing of the epithelial cells (with tight junctions), other mechanisms to keep microbes from gaining entry and colonizing. Skin cells are shed regularly. Ciliated cells of the respiratory tract transfer mucus-entrapped microbes out of the lungs and into the throat where they are first swallowed and then destroyed by stomach acid. The peristaltic movement in the intestine moves potential pathogens out of the body, and the flushing action of the urinary tract is its main defense against potential pathogens.

Because these mechanisms are very effective in removing organisms, a pathogen must adhere to or attach to host cells as a necessary first step in the establishment of infection. As a general rule, both pathogens and host cells are negatively charged and therefore they tend to repel each other. For attachment to occur, the repulsive force must be circumvented.

Some pathogens, such as streptococci that are responsible for tooth decay, possess polysaccharides that form a sticky network of fibrils called the glycocalyx that allows bacteria to attach to surfaces nonspecifically. Other bacteria use pili found on their surface to adhere to the urinary tract (Nester, Roberts, Pearsall, Anderson, and Nester, 1998).

The Teeth

A tooth consists of an exposed crown supported by a neck anchored firmly to the jaw by one or more roots. The roots of teeth fit into sockets. Each socket is lined with a connective tissue periosteum, called the **periodontal** membrane. The root of a tooth is covered with a bone-like material called the cementum. Fibers in the periodontal membrane insert into the cementum and fasten the tooth in its socket. The gingiva, or gums, is the mucous membrane surrounding the alveolar processes in the oral cavity. The bulk of a tooth consists of dentin that is a substance similar to bone, but harder. Covering the dentin on the outside and forming the crown is a tough durable layer of enamel. Enamel is composed primarily of calcium phosphate and is the hardest substance in the body. The central region of the tooth contains the pulp cavity. The pulp is composed of connective tissue with blood vessels, lymph vessels, and nerves. A root canal is continuous with the pulp cavity and opens into the connective tissue surrounding the root through an apical foramen at the root tip

(Van de Graff and Fox, 1995).

Broken Boundaries and Tooth Decays

Although enamel is the hardest substance in the body, it can be weakened by acidic conditions produced by bacterial activity, resulting in dental caries. These caries must be artificially filled because new enamel is not produced after a tooth erupts.

The rate of tooth decay decreases after age 35, but then periodontal diseases may develop. Periodontal diseases result from plaque or tartar buildup at the gum line. This buildup wedges the gum away from the teeth, allowing bacterial infections to develop.

Saliva contains antimicrobial substances,

Table 11.1. Boundaries to Pathogen Entry.

Body Site	Defense Mechanism	Boundaries Broken
Gastrointestinal Tract	0.2% HCl, pepsin	Reduced stomach acid, ingestion of antacid
	Normal Flora	Antibiotics reduce good bacterial numbers; allows for pathogens to replace normal flora
Respiratory Tract	Ciliated cells that constantly move mucous to the throat	Reduced movement of ciliated cells as in smoking, chilling, drugs, bacterial or viral infection, or bacterial exotoxins
Skin	Dryness, acidity, toxicity	Wounds, excess moisture, serous discharge, or because of constant shedding
Teeth	Enamel	Dental caries, periodontal disease
Urinary Tract	Flushing action of urination	Short urethra in women, incomplete urination, sexual intercourse
Vagina	Lactobacilli, acid	Reduced numbers of lactobacilli from douching, soaps, menopause, antibiotic therapy

such as lysozymes, that help protect exposed tooth surfaces. Some protection is also provided by **crevicular fluid**, a tissue exudate that flows into the gingivial crevice and is closer in composition to serum than to saliva. It protects teeth by virtue of both its flushing action and its phagocytic cells and immunoglobulin. Localized acid production within deposits of dental plaque results in a gradual softening of the external enamel. Enamel low in fluoride is more susceptible to the effects of acid.

The dominant microorganisms present are Gram-positive rods and filamentous bacteria. *Streptococcus mutans* is present in small numbers only. *S. mutans* grows well in glucose broth and grows prolifically when sucrose is high in the diet. *S. mutans* is the prime cause of dental caries and broken barriers as they grow in numbers in plaque. Although once considered the cause of dental caries, *Lactobacillus* organisms actually play no role in initiating the process. These very prolific lactic acid producers, however, are important in advancing the front of decay once it is established. A tooth with plaque accumulation is difficult to clean. Decay begins as enamel is attacked by acids formed by bacteria. Decay advances through the enamel and then the dentin. Decay enters the pulp and may form abscesses in the tissues surrounding the root (Tortora, Funk, and Case, 1997).

Skin and Nails

Bacteria and viruses cannot penetrate the tightly knotted cells of the skin. As long as the skin is unbroken, it is one of our most effective defenses. Normal flora (harmless bacteria), such as *Staphylococcus epidermidis*, that live on the skin pro-

tect the body by attacking harmful bacteria that try to take up residence. The acidity of the skin also helps ward off certain harmful bacteria. When bacteria break down the oil on the skin, more acid is produced and the bacteria are prevented from reproducing by their own waste products. Sometimes taking an antibiotic will kill the normal flora on the skin and make it vulnerable to attack by harmful bacteria, fungi, and viruses.

The family of staphylococcus bacteria include the most common inhabitants of the human skin, mouth, nose, and throat. *Staphylococcus epidermidis* is the most numerous skin bacterium and usually causes little harm even when these outer barriers are broken. In contrast, *S. aureus*, a grapelike cluster of Gram-positive cocci, is frequently involved in causing skin diseases. *S. aureus* is a similar bacterium, yet it may cause extensive disease when it penetrates the skin barrier (Alcamo, 1997). Penetration is accomplished by open wounds, damaged hair follicles, ear piercing, dental extractions, and irritation of the skin caused by scratching. Pimples, boils, and carbuncles frequently accompany acne, the skin disease most teenagers and young adults battle.

Mucous Membranes Lining the Body Cavity

The nose and mouth have a sticky coating of mucus that is secreted from the mucous membranes. These mucous membranes consist of an epithelial layer and underlying connective tissue layer. Mucous membranes line the entire gastrointestinal, respiratory, urinary, and reproductive tracts. They form a common internal boundary within the body. Mucus prevents the tracts from drying out. Some pathogens that can survive on the moist secretions of the mucous membrane are able to penetrate the membrane if they are present in sufficient numbers. The penetration may be facilitated by an immune deficiency, toxic substance, prior viral infection, or mucosal irritation. Although mucous membranes do inhibit entrance by bacteria, they offer less protection than skin and nails.

Streptococcus pneumoniae, a normal resident of the nose and throat, is frequently an opportunist. When its numbers are low, they do not cause infection. During times of stress, however, the immune system's defenses are lowered. As these bacteria grow in numbers, an infection can result. When mucous membranes are broken, *S. pneumoniae* can migrate to the lower respiratory tracts causing serious infections. These bacteria once harmless in the nose and throat can become extremely dangerous when they multiply in the lower respiratory tract. When defense barriers are broken, a serious case of pneumonia may result (Nester et al., 1998).

The first sign of disease is an inflammation of the pulmonary tracts. The bacteria may be identified by their cell formation in pairs. *S. pneumoniae* can also be identified on blood agar plates by their alpha-hemolysis (Alcamo, 1997). *S pneumoniae* also produce a dense capsule that makes them resistant to phagocytosis. Pneumococcal pneumonia involves both the bronchi and alveoli. Advanced symptoms include fever, difficult breathing, and chest pain. The lungs have a reddish appearance because blood vessels are dilated. Pneumococci can invade the blood stream, the pleural cavity around the lung, and occa-

sionally the meninges. This is serious pneumonia.

Streptococcus pyogenes, a form similar to *S. pneumoniae,* can also cause serious trouble if mucous barriers in the throat are broken. This pathogen can infect the throat when transferred by air droplets from someone who is sick. Strep throat is an upper respiratory infection caused by beta-hemolytic streptococci. These *S. pyogenes* bacteria are responsible for many skin and soft tissue infections. The pathogenicity of *S. pyogenes* is enhanced by its resistance to phagocytosis. It is able to produce special enzymes that lyse fibrin and produce toxins that are harmful to tissue cells, red blood cells, and protective leukocytes. Strep throat is characterized by inflammation, fever, and tonsillitis where lymph nodes in the neck become enlarged (Alcamo, 1997).

In summary, there are boundaries inside our bodies at the cellular and tissue level. They provide protection against dangerous pathogens and chemicals that might harm us upon entry. We can be good stewards of our bodies by keeping our skin clean and being careful not to touch our nose and mouth with our fingers that may have been contaminated by pathogens. Fortunately, if boundaries of the human body are broken, the Creator has made provision for defenses against infectious disease. These functional body defenses are collectively placed in the immune system.

Our Impressive Immune System
A branch of biology called immunology involves the study of how the body protects itself from agents of disease called patho-

gens, or germs. The word **immune** comes from a Latin root that means "freedom or protection from taxes or burdens." Taken together, we understand that the immune system usually protects us from getting sick and enables us to battle infectious disease successfully. The primary role of the immune system is to recognize pathogens and destroy them. In addition, the immune system prevents the proliferation of mutant cells, such as cancer cells. When the immune system is working properly, it protects the human body from infection. When it is not working or when a boundary is breached, the failure of the immune system can result in localized or systemic infections.

Our immune system acts against specific foreign chemicals and particles based upon prior memory. These foreign particles are called antigens. Antigens include protein-polysaccharide complexes that are part of bacteria, viruses, and protozoans. Once an antigen enters our body, the defense system starts to protect us through an immune response. First, the body must recognize the antigen as a harmful invader and then it must neutralize or destroy it. Both of these steps involve certain types of white blood cells called lymphocytes and neutrophils. Lymphocytes are produced by bone marrow, which is the red tissue in the hollow part of bones (Alcamo, 1997).

Lymphocytes enter the blood and lymph vessels. Cells that mature in the **thymus** gland are called **T** lymphocytes. Others are called **B** lymphocytes because they are processed and mature in the **bone** marrow, or liver. T lymphocytes are further subclassified into helper T lymphocytes and cytotoxic T lymphocytes.

Helper (CD4) T cells act as the directors of the immune system. They identify the enemy and rush to the spleen and lymph nodes, where they stimulate the production of other cells and activate important cells such cytotoxic T cells. Once recruited and activated by helper T cells, cytotoxic T lymphocytes specialize in destroying

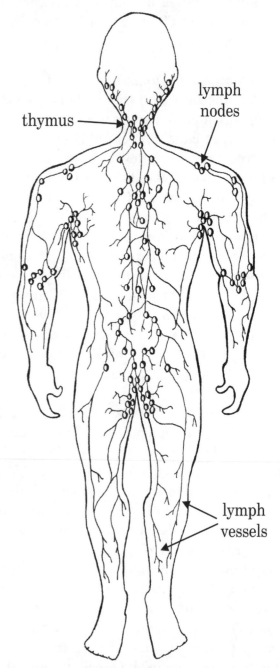

Figure 11.1. Lymphatic system—anatomical defense system.

cells of the body that have been invaded by foreign organisms. They also destroy cells that have turned cancerous.

B lymphocytes act as the "biologic arms factory" because they produce millions of potent chemical weapons called antibodies. They reside in the spleen, or lymph nodes where the helper T cells induce them to replicate. B lymphocytes produce specific antibodies in response to particular pathological agents.

The process continues with macrophages which are large monocytes that are able to "eat" pathogens and also consume other members of the immune system that have been overcome in the struggle. Macrophages clean up the debris after helper-T, killer-T, and B lymphocyte cells have stopped the invaders. Phagocytosis, or "cell eating", occurs when the pseudopods of a macrophage engulf harmful parasites or microorganisms. This is one way that blood serves to cleanse the body of waste and foreign material.

A third type of T cell, the suppressor (CD8) T lymphocyte, is able to slow down or stop the activities of B cells and other T cells. Suppressor T cells play a major role in calling off the attack after an infection has been conquered. This is a simplified version of how our body defenses work at the cellular level. In short, this **"Cell Team"** working in conjunction with natural defense chemicals like interferons, keeps our bodies protected against those outside harmful invaders.

The immune system is somewhat like a protective bubble around the human body shielding us against pathogens and forming an invisible boundary. These vital func-

tions are best illustrated through the case study of a young man who lacked an immune system from birth. David was the first long-term survivor of combined immunodeficiency (SCID) and became known as the "boy in the bubble" or the "bubble boy" through the popular press and television.

David Vetter, the "Bubble Boy"

In 1971, David Vetter was born without a functioning immune system. Faced with what was, back then, a fatal situation, doctors placed the boy in a completely sterile environment. First, David was placed in a small "bubble" crib. They were "buying time" until they could figure out what to do.

Figure 11.2. David Vetter—the "Bubble Boy."

After receiving generous donations, they built him a one-room bubble and after 12 years, David had a two-room bubble and a "space suit" to walk in. The experiment ended in February 1984, when David, the "bubble boy," died. His death provided a clear link between Epstein-Barr virus and cancer. With no "compatible" relatives to provide him with immune cell-producing bone marrow, David had to wait for the technology to rid donated marrow of the cells that would attack his own. After David spent twelve years in the bubble, doctors hoped the technology was ready, and the boy received treated bone marrow from his sister. Eighty days after the transplant and still in a germfree environment, David developed some of the clinical signs of mononucleosis, a condition caused by the Epstein-Barr virus.

Doctors brought him out of isolation for easier treatment, hoping his sister's marrow cells had taken hold and would protect him from the microbes the rest of us encounter every day. Her cells had not established themselves, and David died about four months after the transplant. What killed him was not the immediate failure of the transplant but cancer. David's B cells became disrupted because of the proliferation of the Epstein-Barr virus that caused a rare form of cancer.

The autopsy revealed small, whitish-pink cancer nodules throughout his body, and closer study showed that these cells all contained Epstein-Barr virus—a pathogen that he could have gotten only through his sister's bone marrow. "We're certain (the cancer) came from the transplant itself," says William T. Shearer who was the lead physician on David's case. David had a B cell cancer of a type similar to Burkitt's lymphoma, and while Shearer could not say with certainty the two types of cancer initiated in the same way, "it seems likely that some of these same processes occur" (SoRelle, 1993, p. 1).

David is the most famous case of SCID, the gene for which is carried on the X chromosome. His life was most significant because he was the first longterm survivor, and in his death, a great deal was learned about the immune system.

Ten years after his death, the gene causing SCID was isolated from a cell line that originated with David. SCID usually involves a deficiency of the enzyme adenosine deaminase (ADA). This genetic deficiency leads to a failure of the lymphocytes to develop properly, which in turn causes the failure of the immune system to function (similar to AIDS).

A conclusion that we can draw from David's life is that the immune system acts like a protective bubble around the human body protecting against pathogens. It provides an invisible barrier to localized and systemic infections. Another way of visualizing the immune system is an umbrella protecting our body from the raindrops falling from the sky. When the umbrella is intact, we stay dry; when it breaks we get wet. In the case study of David, his "umbrella" was the bubble and it protected him from "rain" of viruses, bacteria, other pathogens, allergens, and toxins.

This invisible boundary that protects us against pathogens is the result of wondrous interdependent cell types, such as, neutrophils, natural killer cells, T and B lymphocytes, and many others. There are many parts to the immune system, all of

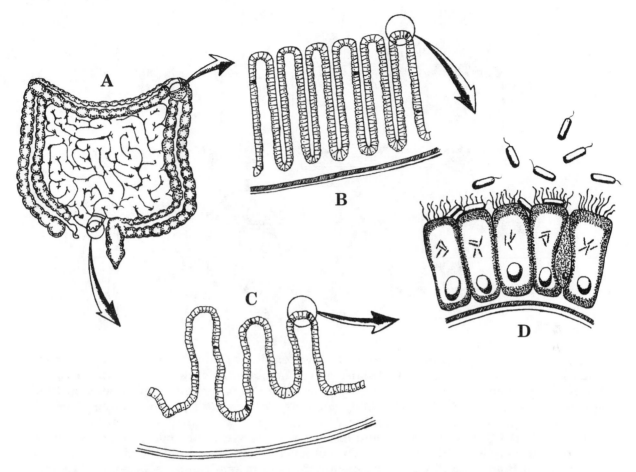

Figure 11.3. Normal flora and boundaries maintained. Most of the time, bacteria (e.g. nonpathogenic *Escherichia coli*) inhabit the gastrointestinal (GI) tract mutualistically. A illustrates the small and large intestines. B is a closeup of the villus lining. C is a close up of the microvillus lining, the site where nutrients are absorbed and transferred to the bloodstream. D illustrates the site where *E. coli* assists in the absorption of micronutrients and manufactures B-vitamins at the surface of the villi.

which must be present for proper defense functioning. It is difficult to conceive how the gradual Darwinian mechanism of mutation and natural selection over billions of years could produce the awesome interaction between the various parts of the immune "cell team."

Final Remarks

We have studied microbial cells that make up the normal flora, like *E. coli* and other enteric bacteria in the intestines. These symbiotic bacteria are usually friendly, but

should not be considered altogether harmless (Brand and Yancey, 1993). Even placid bacteria must not get inside the body or they may become pathogens. Infection from microbes, if left unchecked, can lead to disease and eventually death. Our defensive cells recognize them as foreign and proceed to attack and destroy them. Fortunately, there are epithelial membranes lining the entire gastrointestinal tract and other inside organs that maintain the body's boundaries. These boundaries can also be seen as a marvelous and adaptive mechanism that makes the human body

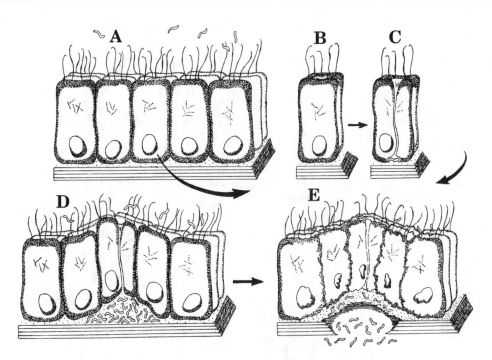

Figure 11.4. Boundaries broken and bacteria invade. The establishment of infection frequently involves an invasion of host tissues when the boundaries are broken. In the lettered sequence, A through E, there is a progressive break in the basement membrane of epithelial tissue because of bacteria, like *Shigella dysenteriae, Salmonella typhi*, or pathogenic *Escherichia coli* strains that release enterotoxins. This poison may disrupt Peyer's patch (a lymphoid tissue in the GI tract) and/or the intestinal wall barrier. This disruption makes room for more bacteria to colonize the tissue. Dysentery, typhoid fever, and bloody diarrhea are possible. In addition, fatal diseases like hemolytic uremic syndrome are possible if toxins from *E. coli* 0157:H7 spread in the bloodstream from the GI tract to the kidney. Injury, disease, and even death are possible when boundaries like these are broken.

unique. These membranes allow needed substances to enter while they exclude potentially damaging chemicals and pathogens. Hence, boundaries are needed for our health and well being. These facts fit quite well with the idea that an all-wise Creator prepared these various boundaries and systems to protect the body from harmful chemicals and pathogenic microbes.

MicroFocus 11.1
The Bible and the Germ Theory

Since his Fall, man has struggled to conquer disease, one of the curses brought upon us by sin. Only in the past two centuries, however, has man made great strides toward curing and preventing disease. Knowledge of the anatomy and physiology grew during the 17th century, but progress in microbiology did not grow very much until the 1850s. In the 1870s, biologists began to link microorganisms with the presence of disease. Pathogens include forms, such as viruses, bacteria, molds, protozoans, rickettsiae, and multicellular parasites.

Louis Pasteur and Robert Koch were able to associate specific bacteria with certain diseases. Pasteur first developed his ideas of the germ theory from his work on fermentation of milk, butter, beets, beer, and wine. Then, Pasteur solved anthrax and rabies disease plagues by treating sheep and dogs using the germ theory concept. Pasteur applied his germ theory to vaccinating people against deadly diseases. Prior to Pasteur, the connection between microorganisms and disease was not apparent since many microbes were known to be beneficial for humans and did not cause disease. Joseph Lister and Robert Koch soon corroborated and extended Pasteur's germ theory. Dr. Lister demonstrated the role of microbes in causing surgical infections and Dr. Koch gave proof that particular bacteria cause specific diseases, such as anthrax and tuberculosis.

During the 1800s, physicians such as Ignaz Semmelweis and Joseph Lister began to understand the role of microbes in causing disease and they applied aseptic principles in their medical practice. Semmelweis washed his hands in solutions of chlorinated lime (a strong disinfectant) prior to delivering babies and the dreaded puerperal fever declined. Lister experimented with a carbolic acid (a phenol), introducing it into wounds during surgery by means of a saturated rag. This technique, along with hand washing, meant that most patients no longer suffered from gas gangrene, septicemia, erysipela, or pyemia after surgery. By the end of the 19th century, patients were less likely to die in the operating room because of disinfected surgical instruments and washing hands. Both Louis Pasteur and Joseph Lister expressed faith in God, and were to some extent "listening" to the Creator to solve medical plagues of their day. The practices of cleansing one's hands, instruments, and other objects date back to ancient Israel when Moses instructed the Israelites about purification. This was not just spiritual cleansing; it also had a useful medical purpose. Long before Semmelweis, Pasteur, Koch, and Lister, the Creator instructed His people in ancient times to distinguish between the unclean and the clean, the antiseptic principle (Leviticus 11:47). In addition, He instructed man, in the Levitical Law, about other disease principles such as quarantine, sanitation of body wastes, and disinfection, long before the Golden Age of Microbiology.

Two thousand years after Moses, medical

practice improved with the development of experimental science, and the establishment of the germ theory of disease. Parts of modern germ theory have historical roots in the Book of Leviticus, where Moses describes the contagious nature of disease. Perhaps modern medicine would have come sooner had physicians "listened" to the laws set down by *Jehovah Rapha*, the Great Physician.

MicroFocus 11.2
Father of Modern Surgery Rejects Darwinism

Joseph Lister (1827–1912) was acquainted with the principle of maintaining boundaries better than anyone alive in his generation. Lord Lister, a devout Quaker physician, is best known for pioneering aseptic surgical methods. Lister is also known as a co-founder of germ theory, as the physician for Queen Victoria, and for working with wound infections. Lister likewise identified himself on numerous occasions as a Bible-believing Christian, believed in special creation and rejected Darwinism.

Lister was a British physician who revolutionized surgery by preventing infection in surgical wounds. Lister, impressed with Pasteur's work on fermentation, (said to be caused by minute organisms suspended in the air), completed his bacteriological studies. In particular, he investigated lactic acid fermentation and milk spoilage. In Lister's paper "On the Lactic Fermentation, and its Bearings on Pathology"(1877), he described his new discovery that a specific bacterium causes the "natural souring" of milk. Through a series of creative experiments, Lister demonstrated that *Streptococcus lactis* was not the result of milk souring, but a cause of milk spoilage as a result of fermentation and putrefaction. Lister thought of milk spoilage as a type of infectious disease. Lister made the connection between how a specific bacterium causes "lactic ferment" in milk and how bacteria in surgical wounds often causes gangrene and pyemia in humans (Gillen and Williams, 1994). Lister's laboratory studies of "milk diseases," in conjunction with Pasteur, led him to study "surgical diseases." Based on his observations of bacteria in air contaminating milk and urine in flasks, Lister was able to apply the germ theory during his operations.

Lister wondered if minute organisms might also be responsible for the pus that forms in surgical wounds. He then experimented with a phenolic compound (carbolic acid) applying it at full strength to wounds by means of a saturated rag. Lister was particularly proud of the fact that after carbolic acid wound dressings became routine for his patients, the patient no longer developed gangrene. His work provided impressive evidence for the germ theory of disease, even though microorganisms specific for various diseases were not identified for another decade. Later, Lister improved his antiseptic surgery to exclude bacteria from wounds by maintaining a clean environment in the operating room and by sterilizing instruments. These procedures were preferable to killing the bacteria after they entered wounds because they avoided the toxic effects of the disinfectant on the wound. His technique of asepsis, along with hand washing, meant that most patients no longer suffered from gas gangrene, septicemia, erysipela, or pyemia after surgery. Lister was knighted in 1883 and subsequently became a baron and a member of the House of Lords. The *British Medical Journal* stated that "he had saved more lives by the introduction of this system than all the wars of the 19th Century together had sacrificed" (Nester, et. al., 1998, p. 202).

Sir Joseph Lister became very popular in his later years. In fact, a popular antiseptic mouthwash, Listerine®, was named in his honor. Early in his life, Lister suffered mild persecution because of his rejection of Darwinism. In one case, he may have been discriminated against in a competitive medical school exam because his ideas did not align with a medical school examiner on Darwin's views of comparative anatomy (Fisher, 1977, p. 54). Later, he contended with Darwinists over abiogenesis. Many Darwinists in the 1880s still believed in spontaneous generation and felt that germs in wounds could magically appear from nowhere (Fisher, 1977, p.181). Like Pasteur, Lister dismissed this an-

cient, disproved idea. Lister (after Pasteur) had to defend his idea of biogenesis that any bacteria found in wounds were a result of contamination in the surgery room. His answer was asepsis, clean hands, and a clean hospital. Perhaps he had read Bible passages like:

"This is the law: ...make a difference between the unclean and the clean..."
Leviticus 11:46–47

and

"Depart ye, go ye out from thence, touch no unclean thing; go ye out of the midst of her; be ye clean, that bear the vessels..."
Isaiah 52:11

In conclusion, Lister revolutionized medicine because he practiced good science. This practice of good science was preceded by a Christian heritage and was grounded in the Bible. In the tradition of Louis Pasteur and Robert Koch, Lister applied the germ theory through his public promotion of cleanliness and sanitation to various governments and institutions all around the world, encouraging them to reduce tuberculosis and other major infectious diseases (Gillen and Williams, 1994). The germ theory of disease is one of the most important concepts that you can understand in the age of AIDS and other deadly infectious diseases. Remember these historical events the next time you are in surgery or use Listerine® mouthwash or toothpaste.

Chapter Twelve
Order, Organization, and Integration

"The brain is a swarm of cells in which everything is seemingly connected to everything else. The connections, though, follow a plan, an order, the large elements of which we are just beginning to see and understand."

Mark Cosgrove, Ph.D. (1987, p. 145)

The brain is the human body's control center. In a sense, the brain is the "head" or chief executive officer (CEO) of the body. Its major parts include the cerebrum, cerebellum, medulla, corpus callosum, hippocampus, hypothalamus, thalamus, and pons. In a typical adult, the brain weighs only a little more than three pounds, yet it consumes about one fourth of the body's total oxygen. Its large demand for oxygen makes it extremely important in terms of bioenergetics.

All the body's cells ultimately report to the brain. In one sense, other body cells are defined by their loyalty to the brain (Brand and Yancey, 1993). They either obey the brain's signals and thereby bring health to the body, or disobey and ultimately bring ruin to the rest of the body.

The human brain is a mysterious loom, weaving strands of 10 billion neurons into the fabric of thought itself (Cosgrove, 1987, p. 145). Perhaps this orderly loom of cellular threads blends with invisible threads of the spirit to form the essence of our personality. There is probably nothing else in the physical universe that is more complex than the human brain. The web of neurons defies description.

The whole mental process consists of neurons transmitting specific chemicals between each other across synapses or gaps. As a result, each cell can communicate with every other cell at lightning speed. In one cubic millimeter (mm^3) of the brain, there are one billion connections among cells. This amounts to about 400 billion synaptic junctions in a gram of brain tissue (Brand and Yancey, 1984). The brain's total number of connections rivals the stars and galaxies of the universe in number, yet the connections follow an orderly

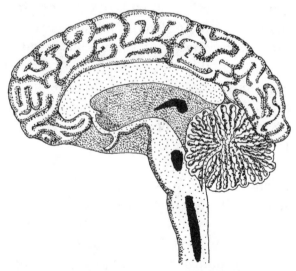

Figure 12.1. Sagittal section of the brain illustrating the reticular formation (dark areas), a complex network that arouses the cerebrum. It integrates various brain signals.

plan.

We can describe physiological interactions in much detail but there is still much unknown about the brain. In one sense, a whole person lies locked inside the cranium, protected, sealed away for managing the duties of the other 70–100 trillion cells in the body (Brand and Yancey, 1984). The brain is the seat of mystery, wisdom, and unity, and it is the source of order for the rest of the body.

Order and Organization Among Neurons

Neurons vary greatly, but they all have the same basic structure. They consist of a cell body and two different kinds of extensions. The first extension is the dendrite. It receives stimuli from its immediate surroundings, from other neurons, or from sensory structures in the skin, muscle, or internal organs. Each cell has many dendrites and each dendrite has many branches. This explains the origin of the word from the Greek *dendron*, meaning "a tree." Because these trees have such a large number of receiving branches, the body of any one neuron is connected to many stimuli sources. A single long extension called the axon (Greek for *"axle"*) carries the impulse away from the cell body and toward other neurons and their dendrites, or directly to muscle and gland cells.

Because the axon is multibranched, it, too, connects with a number of receiving structures. All of this diverse branching serves to enhance integration of the nervous system. A nerve cell may have hundreds of dendrites, but it can have only one axon. The junction at which the impulse is transmitted from one nerve cell to another is

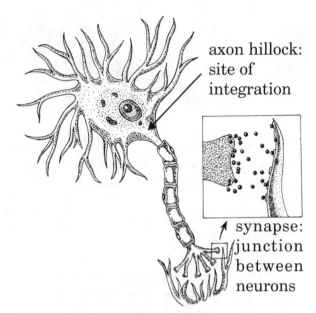

axon hillock: site of integration

synapse: junction between neurons

Figure 12.2. Sites of integration and communication in a typical neuron.

called the synapse, from the Greek *synapto*, "I join together." It is here that the message passes from one of the branches of an axon to one of the branches of a dendrite, or directly to the cell body. The axon to dendrite relationship is reminiscent of Michelangelo's magnificent *The Creation of Adam* on the ceiling of the Vatican's Sistine Chapel. In this painting, the extended right hand of God is reaching toward the outstretched left hand of Adam, but their adjacent fingers are not quite touching. There remains the smallest of spaces between them, across which an impulse must be transmitted (Nuland, 1997).

Michelangelo must have believed that some divine spark of energy leaped across the gap between God and man, and much the same was at first thought about the transmission of nerve impulses through the synapse. But as logical as it seemed, this hypothesis was not supported by later study. When the impulse reaches the tiny intervening space, a chemical substance—

a neurotransmitter—is released from the axon into the junction. In addition to acetylcholine and noradrenaline, there are about fifty other neurotransmitters. The molecules of a neurotransmitter bind to receptor molecules on the dendrite or cell membrane. The binding changes the shape of the receptor; this, in turn, opens up pathways, called channels, in the membrane that envelops the receiving cell.

There is an intricate array of brain neurons (nerve cells) and associated brain cells (for example Schwann Cells). The neuron consists of "processes," known as the dendrites and the axon. These processes transmit signals electrically or chemically to other neurons. Neurons sense electrical impulses by their dendrites and they send integrated impulses through one large axon. The spiderlike dendrites branch out to receive impulses or messages. They are efficient, receiving both excitatory (EPSP) and inhibitory (IPSP) signals from other neurons and body sensors.

Neural Integration

Nerve physiologists define **integration** as a coordination of excitatory and inhibitory signals and impulses received by the cell body and processed at the axon hillock (Campbell, 1996). The nervous impulse must then be integrated at the axon hillock that serves as the neuron's "decision-maker." The hillock acts as a gatekeeper for the axon in that it both integrates the EPSP and IPSP signals and it decides which signals will or will not continue through the axon. Integration requires the processing of diverse electrical signals and produces a whole sum signal that is adaptive, effective, and representative of initial input. The final signal is produced and

sent down the axon. Integration is needed for the sense organs, for perception, for complex nervous responses, and for intelligence.

Other signs of organization seen in the nervous system are the supporting cells for the neurons. These include myelinated glial cells (or neuroglia), Schwann cells, and astrocytes. Schwann cells form the myelin sheaths that cover, protect, and insulate the axon of the neuron. The myelin sheath also speeds up the transmission of ions by allowing signals to "leap" from node to node (Nodes of Ranvier). This phenomenon is known as saltatory jumping.

Another supporting cell designed to regulate tight control over what enters the brain are the astrocytes. These highly integrated cells contribute to the blood-brain barrier and ensure that the brain will not be subjected to chemical fluctuations in the blood. Thus they provide protection against dangerous chemicals and pathogens.

Integration at the Cellular Level

A single neuron may receive information from numerous neighboring neurons via thousands of synapses, some of them excitatory and some of them inhibitory. Excitatory synapses have opposite effects on the membrane potential of the postsynaptic cell. At an excitatory synapse, neurotransmitter receptors control a type of gated channel that allows Na^+ to enter the cell and K^+ to leave the cell. There is a depolarization in the cell. Therefore, the electric change is caused by the binding of the neurotransmitter to the receptor. Those signals that collectively lead to a positive action potential down an axon are referred to as excitatory postsynaptic potentials (EPSP). It should be noted that it takes a minimum number of EPSPs to stimulate a signal strong enough to initiate communication along the axon. This pattern was discussed previously as the **all-or-none principle**.

At an inhibitory synapse, the binding of neurotransmitter molecules to the postsynaptic membrane hyperpolarizes the membrane by opening ion channels that make the membrane more permeable to K^+ or to Cl^-, which rush out of the cell. Therefore, the voltage change caused by the chemical signaling to the receptor is called an inhibitory postsynaptic potential (IPSP). The sum of the EPSPs and the IPSPs are integrated at the axon hillock and an intelligent signal is sent down the axon (Campbell, 1996).

This complex integration works much like the computer chips in your personal computer. Paul Brand and others have tried to copy this mechanism when building an artificial pain system (Brand and Yancey, 1993). They achieved some success with their master engineers, electricians, and nerve physiologists in producing temporary pain receptors and signals. After spending over a million dollars of grant

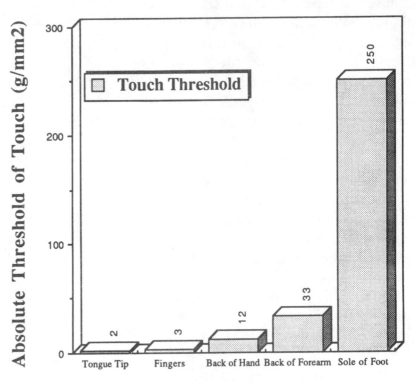

Figure 12.3. The minimum pressure needed to feel touch in various parts of the body (after Brand and Yancey, 1993).

money, however, they basically failed to make a complete copy of this detailed integration of the nervous system. Their conclusion was that this integration of sensory receptors for pain can be viewed only as the fingerprint of a wise master electrician. Indeed there is a Bioengineer behind the intricate integration of the nervous system.

Integration at the Network Level

Integration can be noted at the neuron level. The sensation of pain is a good example. Millions of touch sensors over the surface of the skin, are scattered nonrandomly, but in precise accord with the body's specific needs. The body does not seem to have any special cells dedicated to the sensation of pain because pain is tied to an elaborate network of sensors that report degrees of pressure, touch, heat, and cold.

Scientists blindfolded their research subjects and measured their skin sensitivity (Yancey, 1990). They studied how much pressure must be applied before a blindfolded person becomes aware of an object touching his skin. Although the skin is a single organ, it displays a wide range of sensitivity to pressure. The skin must be able to sense pressure in such diverse tasks as picking food particles from our teeth, using fingers to play the guitar, or writing with a felt tip pen. These areas of the skin require a high degree of sensitivity.

But less critical areas hardly need such sensitivity. Indeed, we would tire very quickly indeed if our brains had to listen to such dainty pressure reports from the foot, which faces a daily rigor of stomping, squeezing, and supporting weight. Thus, while fingers and the tongue can detect a feather touch, other parts of the body need a good sound slap before they report unusual activity to the brain (Brand and Yancey, 1993).

These measurements of touch threshold barely scratch the surface of the marvels of the pain network. For example, sensitivity to pressure varies depending on context. Humans can distinguish a letter that weighs ¼ ounce from one that weighs ½ ounce just by holding it in their hand (Brand and Yancey, 1993; Yancey, 1990). But if they try to distinguish a 10 lb. package from a 10.3 lb. one, they cannot discern the difference. Our sensitivity to pressure is amazing, yet it does have limits.

Pain Network Is Ordered

Another test assesses the *absolute threshold of pain*. In this test, the scientist measures how much pressure must be applied to a very sharp needle before the subject begins to experience pain. The fingertip, for example, shows an astounding difference; it can detect a mere 3

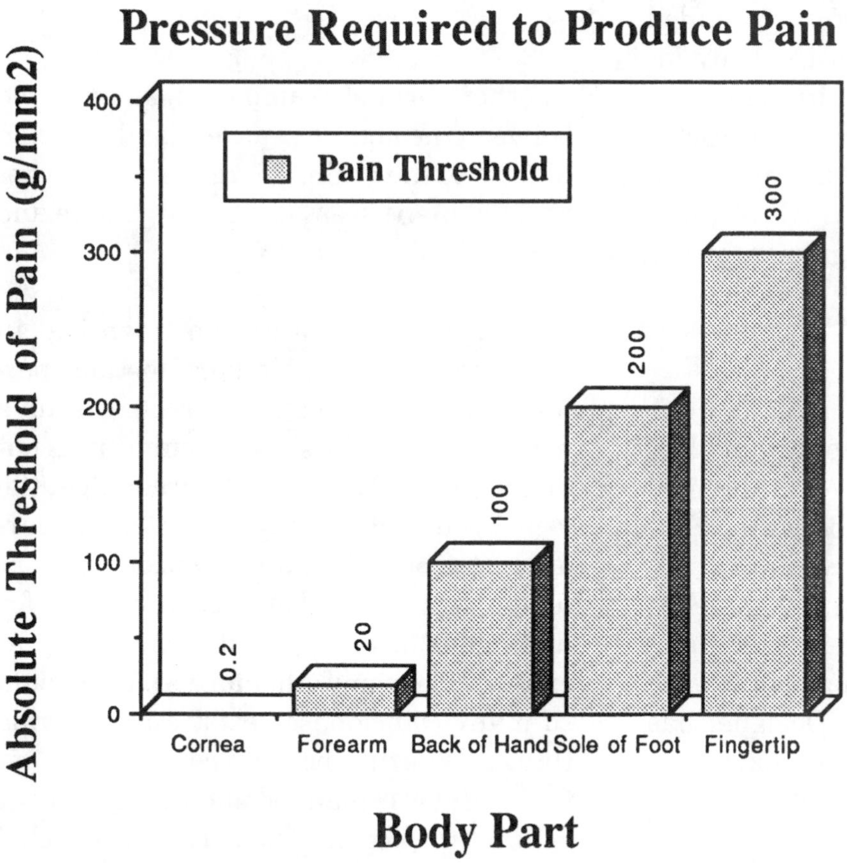

Pressure Required to Produce Pain

Figure 12.4. The minimum pressure needed to produce pain in various parts of the body (after Brand and Yancey, 1993).

grams/millimeter^2 (g/mm^2) of pressure, but pain will not be felt until that pressure exceeds 300 g/mm^2 (Brand and Yancey, 1993; Yancey, 1990). Why? The concert violinist must sense an amazing range of pressures to produce perfect sound and volume. A skilled pizza maker, swishing his hands through batches of dough, can notice as little as a two percent variance in its "stickiness" consistency.

The fingertips must be incredibly sensitive to the slightest differences in touch. But sensitivity to touch is not enough. The fingertips must also be able to withstand rigorous activity. Next time you meet a carpenter or an experienced tennis player feel their calloused hand. Life would be miserable indeed if the fingertip fired a message of pain to the brain each time a person gripped a tennis racket or pounded a hammer. So the design of the body includes fingers that are sensitive to pressure, but relatively insensitive to pricks. Hands and fingertips serve us well, as do the other parts of our body. All of these require sophisticated integration of the pressure network.

Parallel Plans

In the human body, we see order and levels of organization. In fact, we can see parallel or congruent patterns between cellular organelles and the body organs. Although the comparison between organelles to organs is not exact, the similar functions performed by each corroborate the principle that an intelligent Designer has made both. We conclude that mankind has been fashioned by a wise Creator.

Parallel plans can be observed at different levels of organization. This congruence is most notable in a comparison between organelle and organ (Table 12.1). The chart compares the functions of organs to the functions of organelles. The concept implied by the chart is one suggesting that a Master Bioengineer used a similar scheme. Engineers use the same general mechanical principles in the manufacture of gears regardless of whether the gears are for an automobile or a watch (Kaufman, 1995).

Examples of this principle are many. The cell membrane can be compared to the human body's skin. One is the outer covering that holds the body together and the other holds the cell together. Eleven organs in all are compared to 11 similar organelles in order to provide evidence for this concept of parallel plans.

Another parallel example can be observed between the neuron's nucleus and the axon hillock, to the brain. There are many other parallel plans seen between organelle and organ (Table 12.1).

Although the parts are not identical in structure, there are similar functions performed. Like a computer chip and a motherboard in PCs, where circuitry is not directly interchangeable between PCs, the logic used in designing both plans is parallel. It took order, organization, and intelligence to make both. The evidence for intelligent design is solid and highly unexplainable in any other format. How could so many coincidences exist among living things? Was this merely a coincidence that this world was planned and thought about before time began? The perfectly modeled parallel plans were made by an all-wise Inventor.

Table 12.1. Organelles Compared to Organs. Modified from Kaufmann (1995, p. 241) with permission.

Organelles and Their Functions	Organs and Their Functions
1. Cell or plasma membrane—envelops cytoplasm allowing inflow and outflow of materials to or away from the cytoplasm.	Skin, and outer covering of the epithelium and connective tissue controlling a limited inflow and outflow of materials to or away from the body.
2. Endoplasmic reticulum (ER)—sheets of membrane (rough and smooth), involved in protein synthesis and detoxification of lipid-soluble drugs.	Liver cells that process proteins and detoxify lipid-soluble drugs (alcohol and phenobarbital)
3. Ribosomes—small structures containing RNA and protein that synthesize other proteins. Some are free; others attach to ER making it rough ER.	All cells since they build and maintain the protein components of their structures.
4. Golgi apparatus—membranous vesicles stacked like hollow saucers that package and secrete proteins and membranes made by the rough ER.	Cells of endocrine and exocrine glands especially mucous cells of the digestive tract that secrete mucus.
5. Mitochondria—double membrane-bound structures which are the main site of oxidative operations and production of ATP the fuel for muscular contraction.	Muscle cells—contractile units that split ATP which allows contraction by mechanical work.
6. Lysosomes—membranous sacs that secrete enzymes which disintegrate harmful biological molecules.	Phagocytic cells (lymphocytes) that degrade ingested bacteria and viruses.
7. Peroxisomes—membrane-lined sacs resembling small lysosomes that secrete enzymes that range from dangerous free radicals to harmless water and oxygen.	Liver and kidney cells which perform similar detoxification processes.
8. Microtubules—hollow tubes providing structural support in the center of the cell.	Bones of the axial skeleton (head, shoulder girdle, vertebral column, pelvic girdle) that give structure to the center axis of the body.
9. Microfilaments—strands of protein (actin and myosin) that slide inward causing a sarcomere (unit of contraction inside muscle cell) to shorten.	Bones of the appendicular skeleton (upper and lower extremities) which support the outer body parts.
10. Intermediate filaments—microfibers with a diameter between microtubules and microfilaments that resist forces in the outer part of the cell.	Muscle cells that shorten.
11. Nucleus—large spherical structure containing nucleolus and genetic material (DNA) that controls the cell's activities.	Brain (cerebral hemispheres, diencephalon, brain stem and cerebellum) which activates and controls all cells and organs.

Summary and Conclusions

The biological basis for intelligence in man or animals is both controversial and uncertain in the world of secular science. Yet, the complex integration of information through various neurons in the cerebellum is most likely the biological basis for intelligence. Although a precise integration "circuit" is probably not known, one can confidently infer that intelligence itself is an outcome of integration. At the physiological level, a behavior pattern is the action of an animal's effector in response to a stimulus detected by its receptors. In between receptor and effector lies the nervous system that determines what information travels from one part to the other. The nervous network, intelligence, and behavior are still poorly understood.

Neural inhibition is the capacity of a neuron to exert a fundamental property of the nervous system. An old-fashioned telephone exchange worked not so much by connecting a caller to the person with whom he wishes to speak, but by keeping the caller from getting a wrong number. In making a behavioral decision, neural networks are designed to be able to inhibit all systems except the one responsible for ordering the desired decision. If a nervous system were unable to do this, the result would be neural chaos, behavioral paralysis, and death.

Intelligence is probably a coordinated effort by the nervous network to integrate and process signals appropriate for the particular circumstance. The completion of an intelligent activity by man (or an animal) may stem from a rigidly patterned neural network. The network must gather and process data from key neurons in the cerebrum. Then it must order particular sequences elsewhere in the body while blocking signals that might inhibit an effective behavior. A comparison among neurons, neural networks, and the brain indicates the parallel plans of integration. It causes us to think that there is a Higher Intelligence behind the design of this intricate nerve integration. Can one explain design, integration, and intelligence without a Master Mind?

Recall that the theory of macroevolution predicts randomness, blind chance, imperfection, expansion of the gene pool, and tinkering in nature. In contrast, a creation model predicts order, organization, boundaries to the gene pool, and integration. The nervous system has truly amazing properties. Do the properties of order, organization, and integration found in the nervous system correlate better with the characteristics of macroevolution or intelligent design?

Order and organization are two principles that are consistent with a Creator. Webster defines **order** as a condition in which everything is in its right place and functions properly. In addition, order implies there is a fixed or definite plan and system. Webster defines **organization** as a unified, coherent group or systemized whole. A divine craftsman would put things into order from the beginning, as a master carpenter would have his blueprint and materials ready for construction ahead of time. Based upon the complexity of this integration, it seems that only the Greatest Mind in the Universe could have accomplished all of this.

Chapter Thirteen

Patterns in Physiology

"For the life of the flesh is in the blood."

Leviticus 17:11

Throughout this book, we have explored the basic principles in physiology of the human body. These patterns can be observed in all eleven of the human body systems. They include 1) the relationship of structure to function; 2) homeostasis; 3) interdependence among body parts; 4) short-term physiological adaptation; 5) maintenance of boundaries; and 6) the triple theme of order, organization, and integration. These themes are consistent with a design or creation perspective of the human body.

This organization is different from most anatomy and physiology (A and P) textbooks or other books on the human body. Many A and P books use a systems approach, where most chapters are devoted to one of the eleven body systems, such as the circulatory system. Some medical A and P texts use a regional approach, where one body region, such as the thoracic region, is treated at a time. This book is somewhat unusual in its organization around the physiological themes and in its creationist approach to the human body. The physiological patterns are almost universally recognized by professional biologists, but seldom discussed chapter by chapter. We believe that if you will spend time with both a traditional A and P text and this one, your understanding of the human body will grow.

Also, recall our theme that the human body is like a machine in many ways. Each part of the body has its own job. The body and a machine both perform work. This analogy is still applicable today. In addition, we have seen how anatomy and physiology themes are consistent with a creation perspective of the human body. Many of the details discussed in Chapters 2 through 11 would be predicted by a design view of the human body.

Life Is in the Blood

Blood is a rich scarlet soup of proteins and cells that keeps us alive. A few nights working in an emergency room would probably convince you that the body is just a huge bag of blood. Actually, in an average 70 liter human body, only 5 liters, or 7% by volume, is blood. Normally blood is found in the heart; in blood vessels; and in the sinusoids of the marrow, liver, and spleen.

Of the average five liters (l) of blood, only 2.25 l, or 45%, consists of cells. Erythrocytes, leukocytes, and platelets are the **formed elements** of blood. The rest is plasma that consists of 91.5% water (by weight) and 8.5% solids (mostly albumin). Of the 2.25 l of cells, only 0.037 l (1.6%) are leukocytes. The entire circulating leukocyte population, if purified, would fit in a coffee cup. The total circulating platelet volume is even less, about 0.0065 l, or about one teaspoonful.

Because blood is the connecting fluid to all the body systems and is the substance upon which all body cells are dependent, we explore blood as it relates to all the physiological patterns. These are universal patterns that almost all physiologists would recognize. In this section, we discuss the biological basis of blood and how it relates to these themes.

Blood and Homeostasis

Recall Bernard's idea of body balance known as the "constant internal milieu." He recognized the power of many simple animals to maintain a relatively constant condition in their internal chemical environment even when the external chemical surroundings change. An ocean-dwelling jellyfish is powerless to affect the water, salinity, and electrolytes of the sea in which its cells dwell, but the jellyfish can maintain its "internal pond" temperature and the correct osmolarity. Likewise we

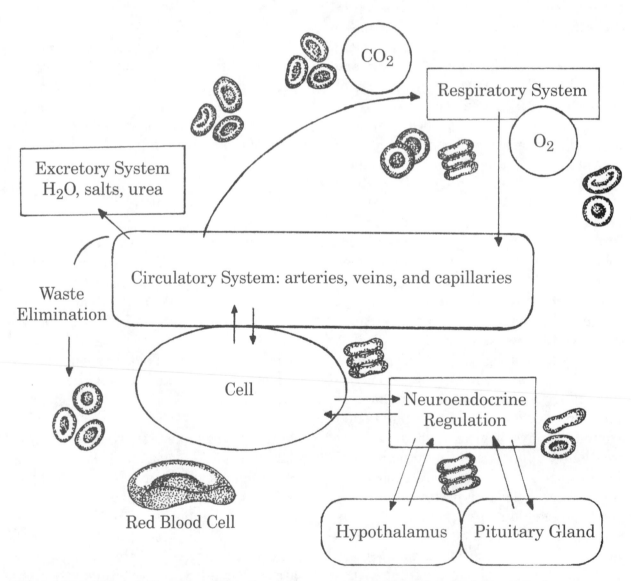

Figure 13.1. The relationship of the circulatory system to the other body systems in maintaining homeostasis (regulation of the body).

see this idea lays the foundation for understanding the regulation of substances in the blood. Figure 13.1 shows the relationship of the circulatory system to the other body systems in maintaining **homeostasis.** Notice how oxygen, carbon dioxide, nutrients, hormones, and immune substances are transported by the blood to various parts of the body.

Blood Compared with Seawater

Blood resembles seawater in many respects. The components of ocean water and blood have much in common in terms of chemical and physical characteristics (Table 13.1). Both blood plasma and seawater carry electrolytes, such as Na^+, K^+, Ca^{+2}, Mg^{+2}, Cl^-, HPO_4^{-2}, SO_4^{-2}, and HCO_3^-. In addition, they have substances in common that are transported routinely in their "currents." They also carry nutrients, regulatory substances, gases, and wastes. Blood and our interstitial fluid serve as an internal ocean for the body's cells.

On the other hand, there are some key differences between blood and seawater. Blood is heavier, thicker, and more viscous than seawater. Consequently, it flows more slowly than water, at least in part because of its viscosity. Blood likewise differs from seawater in that it has an adhesive quality, or "stickiness" to it. This trait may be appreciated by touching blood. Furthermore, blood plasma is different from seawater in that it carries proteins, such as albumins, globulins, and fibrinogen (about 7% of plasma is protein). Although these two fluids are not the same, it is helpful to use the ocean and sea life as an analogy for blood (and its formed elements) when studying homeostasis.

RBC Form and Function

A study of erythrocytes or red blood cells (RBCs) illustrates the correlation between structure and function. RBCs are structurally the simplest cells in the body. RBCs are highly specialized for their oxygen transport function. Each one contains an amazing 280 million hemoglobin molecules. Although red blood cells initially have a nucleus, they lose it as they mature (Fig. 13.2).

Since RBCs do not have nuclei, all their internal space is available for oxygen transport. The shape of the RBC facilitates its function. A biconcave disc has a much greater surface area for its volume than a sphere or a cube. The shape confers two advantages. First there is a large surface area for the diffusion of gas molecules in

Table 13.1. Comparison of physical and chemical properties of seawater and plasma (data compiled from Van de Graff, 1998, p. 524).

	Seawater	Plasma
electrolytes	numerous, diverse	abundant, diverse
pH	7.6–8.3	7.35–7.45
salts and solutes	3.5%	8.5%
relative viscosity	1–2	4.5–5.5
water	96.5%	91.5 %

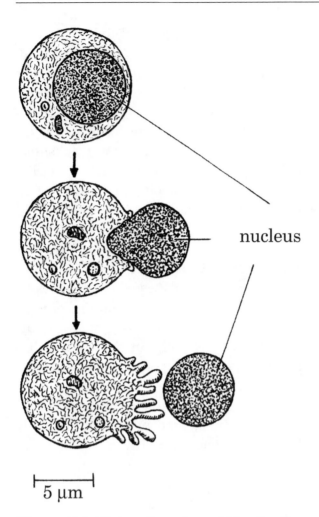

nucleus

5 µm

Figure 13.2. Maturation of a red blood cell.

or out of the RBC. Second, the biconcave disc is very flexible which permits RBCs to squeeze through small capillaries, some of which are as narrow as 3 µm.

The basic function of the RBC is the creation and maintenance of an environment salutary to the physical integrity and functionality of hemoglobin. In the normal state, erythrocytes are produced only in the bone marrow of the skeleton (in adults only in the axial skeleton), but in pathologic states (especially myelofibrosis) almost any organ can become the site of erythropoiesis. Numerous substances are necessary for the creation of erythrocytes, including metals (iron, cobalt, manganese),

vitamins (B_{12}, B_6, C, E, folate, riboflavin, pantothenic acid, thiamine), and amino acids. Regulatory substances necessary for normal erythropoiesis include erythropoietin, thyroid hormones, and androgens.

Erythrocytes arise from blast precursors in the marrow over a period of five days. Then they are released into the blood as reticulocytes, distinguishable from regular erythrocytes only with special stains. The reticulocyte changes to an erythrocyte in one day and circulates for 120 days before being destroyed in the reticuloendothelial system.

Leukocytes Help Maintain Boundaries

Blood is also the body's cleanser. Several leukocytes (or white blood cells) illustrate the principle of protection by maintaining boundaries. Leukocytes are manufactured for whatever defense is needed. Dr. Ronald Glasser considers "this process a mixture of mystery and chemistry, as well as a combination of physics and grace down at the molecular level." Glasser gives the impression that he is amazed by the precision in which leukocytes operate, yet he does not have a sufficient explanation for their origin. He simply attributes to leukocytes a new level of evolutionary sophistication (Glasser, 1976, p. 70). Leukocytes actually fulfill a creation prediction that the human body was made with certain lasting boundaries for its long-term survival.

Neutrophils are the most populous of the circulating leukocytes. They are also the most short-lived in circulation. After production and release by the marrow, these white cells circulate for only about eight hours before proceeding to the tissues. They live in these tissues for about a week if all goes well. They are produced in response to acute body stress, whether from infection, infarction, trauma, emotional distress, or other noxious stimuli. When neutrophils are called to an injury site, they phagocytize invaders and other undesirable substances. They usually kill themselves in the process of destroying pathogens.

Neutrophils are the most numerous type of leukocyte. Frequently, the circulating neutrophil series is an indicator of stress or disease in the body. Their numbers, types, and band counts are used as an indicator of acute stress. They are also a sign that boundaries have been broken.

Platelets Illustrate Interdependence

Recall that platelets and other cascading components of blood clotting illustrate the theme of interdependence. Here we discuss what happens if platelets and their cascading components do not work interdependently. If macroevolution were true, then there would be times when one or more of the interdependent chemicals have been absent in the piecemeal evolution toward a blood clotting mechanism. In today's world, individuals who lack one or more of the interdependent, coagulation chemicals are referred to as having **hemophilia**. In this section, we explore hemophilia and its consequences.

The term "hemophilia" actually refers to several different hereditary deficiencies involving coagulation. The effects of each disorder are so similar to each other, they are hardly distinguishable. But each is a deficiency of a particular blood-clotting factor. Hemophilia is characterized by subcutaneous and intramuscular hemorrhaging, nosebleeds, blood in the urine, joint pain, and damage due to joint hemorrhag-

ing. Treatment includes applying pressure to accessible bleeding sites and transfusing fresh plasma or the appropriate deficient clotting factor to relieve the bleeding.

The most common form of hemophilia occurs because of a lack of an antihemophilic factor which in turn helps the Christmas factor in the conversion of the Stuart factor to its active form. Lack of the Christmas factor is the second most common form of hemophilia. Severe health problems or even death can result if defects occur in other proteins of the clotting pathway.

Essentially every form of hemophilia could have occurred during evolution of a blood clot if this gradual mechanism of natural selection on chance mutations were true, as Darwinists contend. How could any hemophiliac trait have survival value? This trait does not add to any biological fitness measure. Therefore, we must conclude that platelets and their "all or none" components in blood coagulation are necessary parts of a molecular team. Most logically, these formed elements are the work of a Creator.

Effects of High Altitude on Blood

Imagine yourself training for mountain climbing. Your plan is not to climb just any mountain, but to ascend to the top of Mt. Everest. In order to move up to this high altitude, your body will have to undergo a number of short-term adaptations, most of them related to your cardiovascular system. In particular, your heart, lungs, and blood will have to change in order to accomplish this incredible feat. During acclimatization at high altitudes, there is an increase in oxygen carrying capacity.

Recall from Chapter 9 that hypoxia (low oxygen availability to the body) is the principal stimulus for causing an increase in red blood cell production. Ordinarily, in full acclimatization to low oxygen, the **hematocrit** (a measure of the number of red blood cells) rises from a normal value of 40 to 45 to an average of 60 to 65, with an average increase in hemoglobin concentration from a normal of 15 gm/dL to about 22 gm/dL (dL = deciliters, tenths of a liter). In addition, the blood volume increases, often by as much as 20 to 30 percent, resulting in a total increase in circulating hemoglobin of as much as 50 to 90 per cent. Unfortunately, this increase in hemoglobin and blood volume is a slow one, having almost no effect until after 2 to 3 weeks, reaching half development in a month or so, and becoming fully developed only after many months. The longer you train for mountain climbing, the more your body can handle the rarefied air and vigorous aerobic exercise. There are limits to how far your body will acclimatize.

Some factors that affect this adaptation at high altitudes include 1) an increase in pulmonary ventilation; 2) an increase in diffusing capacity for oxygen in the lungs; 3) an increase in hemoglobin of the blood; 4) an increase in capillarity of the circulatory system (greater capillary density); 5) greater vascularization in the body as a whole; 6) an increase in the ability for cells to utilize oxygen despite low pO_2. (Number 6 takes place by cell acclimatization, whereby mitochondria and some cellular oxidative enzyme systems become slightly more plentiful.); and 7) an increase in size of the right side of the heart to handle the increase in pulmonary circulation.

It is the sum effect of all the systems adapting that allows for long-term survival at high altitudes; this is why native mountain-dwellers can climb to the top of Mt. Everest without much assistance from artificial oxygen. In contrast, normal lowlanders and even trained mountain climbers cannot ascend Mt. Everest without some help from an artificial oxygen tank. Some of this adaptation occurs from birth and must occur in early life. Usually, it is not altogether attainable in later life. Capillary expansion is developed in early life. This does not occur in adults even if they have lived there for ten years or more. Native Peruvians in the Andes have increased heart and lung capacity developed from birth.

Further proof of the importance of natural acclimatization of natives is that the Sherpas of the Himalayas can survive without artificial oxygen for short periods of time (hours) at altitudes as high as the top of Mt. Everest—over 29,000 feet.

These native mountain people are generally short in stature and have broad chests. Their chest cavity, in proportion to total body size, is greatly increased, to give a high ratio of ventilator capacity to body mass. In particular, the right ventricles of their hearts are larger to achieve a high pulmonary arterial pressure. This increased pressure is a compensation for pumping blood through a greatly expanded pulmonary-capillary system.

One beneficial effect of this adaptation for these natives is an increased work capacity at high altitudes. In general, there is an increase in maximum rate of oxygen uptake that the body can achieve. The work capacities in percent of normal for unacclimatized and acclimatized persons

Unacclimatized	50%
Acclimatized for two months	68%
Native living at 13,200 ft., but working at 17,000 ft.	87%

Table 13.2. Residents from Different Altitudes and Their Work Capacity at High Altitudes.

at an altitude of 17,000 feet are shown in Table 13.2 (Guyton, 1991, p. 466). Thus, naturally-adapted natives can achieve a daily work output at this high altitude almost equal to that of a normal person at sea level, but even when acclimatized, lowlanders can almost never achieve this result. In Figure 13.3, observe the physiological differences between three groups of people and their bodies' adaptation to pick up oxygen in their environment. Once again, we observe that blood is key to adaptation.

Lymphocytes and Organization

Lymphocytes, a specific type of leukocyte, are extremely diverse and their system of

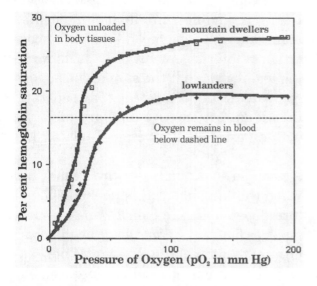

Figure 13.3. Oxyhemoglobin dissociation curves for residents from different altitudes. Oxygen loads in body tissues above dashed line. Data from Guyton, 1991.

communication illustrates the triple theme of order, organization, and integration in the body. In the immune/inflammatory response, if the neutrophils and monocytes are the brutes, the lymphocytes are the brains. It is possible to observe the horror of life without lymphocyte function by studying the unfortunate few with hereditary, X-linked, severe combined immune deficiency.

Such individuals uniformly die of systemic infections at an early age (except for the "bubble boys" of yesteryear, who lived out their short lives in antiseptic rooms). The functions of lymphocytes are so diverse and complex that they are beyond the scope of this text. What follows are a few general remarks concerning their role in the order, organization, and integration of the immune system. After neutrophils, lymphocytes are the most numerous of the circulating leukocytes. Their life span may vary from several days to a lifetime (as for memory lymphocytes).

Recall the discussion of the immune response in Chapter 11 (Boundaries) and how the helper T-cell acts as commander-in-chief of the immune system. It identifies the enemy and rushes to the spleen and lymph nodes, where helper T-cells stimulates the production of other cells, such as killer T-cells. Helper T-cells are also referred to as CD4 cells. CD4 cells organize how the killer T-cells are recruited and activated. They also activate B-cells which act as the biological arms factory in producing antibodies. In addition they summon macrophages that "clean up the mess" after helper T, killer T, and B-cells have stopped the enemy. Phagocytosis, or "cell eating" occurs when the pseudopods of a macrophage engulf a number of bacteria or other parasites into itself.

The lack of this organization leads to immunodeficiency. Today, the most common type is acquired immunodeficiency syndrome (AIDS). Many think of AIDS as an acute infection by HIV. Its first signs are swollen lymph nodes. (Frequently, people with an acute HIV infection experience fever, diarrhea, rash, swollen lymph glands, night sweats, and a general fatigue similar to that in acute infectious mononucleosis). It should be pointed out however, that the Center for Disease Control defines AIDS as occurring when the **CD4 cell** count in blood is *less than 200/µl* (millionths of a liter). In this way we see the importance of CD4 in organizing and communicating with the rest of the immune system.

Blood: First to Live and Last to Die

In blood, we study the reason for our existence. Blood is one substance that the human body depends upon for life. William Harvey not only revolutionized our understanding of the heart and circulation, but he also illuminated our knowledge of blood. Harvey said:

Blood is the cause not only of life in general but also of longer or short life, of sleep and watching, of genius, aptitude and strength. It is the first to live and the last to die.

Brand and Yancey, 1984, p. 66

Blood signifies life and not death to the medical practitioner. It feeds every cell in the body with its precious nutrients. When it slips away, our body falters.

Interdependence

In many respects, the theme of interdependence is the strongest evidence for creation. The concept of interdependence of body parts is one of the most recognizable themes in the disciplines of biochemistry, cell biology, and human anatomy and physiology. It is repeated in many diverse sets of cells, organs, and organ systems.

The famous comparative anatomist George Cuvier, in the 18th century developed a similar notion of irreducible complexity, calling it the **correlation of parts**. Cuvier, a firm creationist, advanced this argument in debates against evolution (Morris, 1988). In the 20th century, other scientists, such as Dean Kenyon, Michael Denton, and Michael Behe have also argued for design in light of the evidence of the interdependence and complexity of body structures.

Related terms used by various authors to describe this phenomenon of interdependence include adaptational package, cell team, compound traits, emergent properties, irreducible complexity, molecular team, and synergism. This resulting condition of interdependent body parts working together is that the sum of the actions is greater than the addition of all separate, individual actions. A summary of terms with an example is given in Table 13.3.

A common analogous structure we may use when thinking about this phenomenon of interdependence is a mousetrap. A mousetrap has five parts: a platform, holding bar, hammer, catch, and spring. When assembled there is no gradual improvement of function. It does not work until every part is in place. The same thing is true inside a living cell, as well as in specific

Table 13.3. Interdependence in the human body.

Term	Example
Adaptational package	*E. coli* and digestive system
Cell team	Immune system
Compound traits	The eye and its muscles
Emergent properties	Liver cells
Interdependent parts	Excretory system
Irreducible complexity	The hand
Molecular team	Blood clotting
Synergism	The body as a whole

organs in the human body. Many of its systems will not work unless every part is there at the same time. There is a tremendous advantage to cooperation in the body at all levels from the molecular to the systematic level.

In summary, we have discussed more than a dozen examples of interdependence. These are but a few of the many irreducibly complex systems in the human body. Other systems that illustrate this concept of interdependent, anatomical compound traits are the many lever and pulley mechanisms found among the skeletal and muscular systems in the body. The muscles, cartilage, ligaments, bones, tendons, and articulations (joints) interacting with each other frequently work like pulleys to move the body. The pulley is one of man's most work-saving machines. For a discussion of how pulley and lever systems are intricate systems in the human body, see Kaufmann (1980; 1981; 1994).

Kaufmann argues that these pulleys, like other simple machines, have interdependent parts, implying an intricate design.

It is logical that these complex, mechanical systems did not evolve by consecutive random accidents of physiochemical forces, but are the result of planned thinking by a Divine Workman. Therefore, we must conclude that a Creator has formed and fashioned the human body.

The Ultimate Machine

We can compare the workings of the human body with an automobile. This discussion of man as *"the ultimate machine"* is modified from an advanced physiology lecture by David Kaufmann, (1997). Our brilliant physical scientists have produced all kinds of efficient inanimate machines, but the greatest inventor of all, the Triune God of Creation, has invented the greatest machine ever, the human body. Although the human body cannot be totally explained in terms of mechanistic or machine analogies, we return to the comparison between man and machine.

The dictionary definition of a machine is an apparatus consisting of interrelated parts with separate functions that jointly

Table 13.4. Comparison of the automotive machine to the human machine.

Auto Systems	Human Body Systems
Electrical (stored energy)	Bone (skeletal framework)
Ignition (starts energy flowing)	Joint (articulations)
Combustion (explodes, mechanical work)	Muscle (pulling power)
Transmission (levels of efficiency, gears)	Nervous (electrical)
Cooling (keeps from overheating)	Endocrine (chemical controller)
Front end (steering)	Cardiovascular (motor)
Rear end (rolling power)	Lymph (immunity)
Brake (safe stopping)	Respiratory (gas exchange)
AC/Heater (environmental control)	Digestive (nutrition/excretion)
Exhaust (emits harmful gases)	Excretory (liquid waste)
Assembly Line (new cars produced)	Reproductive (regeneration of cells and organisms)
Paint (protects and beautifies car)	Integumentary (covers and protects car)

perform work. Table 13.4 makes a comparison between the eleven systems of the automobile and the eleven systems of the human body. For example, the skin protects the body as paint protects the metal on a car. Food serves as fuel for the body as gasoline does for a car. Like a machine, the human body wears out and breaks down when not properly maintained. Both machines are well-manufactured, function efficiently, and serve us superbly. The human machine is far superior, however, because its Designer and Manufacturer is The Lord Jehovah.

The Superiority of the Human Body as a Machine

Although the automobile has many features in common with our body, clearly the human machine is more complex. First, the human body uses a variety of fuels, including carbohydrates, fats, and even proteins sometimes. The automobile uses only one fuel, gasoline.

Second, the human body excretes only recyclable (biodegradable) wastes (H_2O, CO_2, sweat, urine, and feces). The exhaust from the automobile is not biodegradable, but is biodevastating, contributing to many types of degenerative diseases. Third, the human body accepts a wide range of informational inputs which we call sensations, such as sight, sound, taste, smell, balance, pressure, touch, pain, and even movement. The automobile is limited to only such basic environmental inputs as wind, rain, dust, and friction.

Fourth, the human body performs amazing feats of data processing, through the functioning of the central and peripheral nervous systems involving conductivity, synapses, reflexes, circuits, pathways, and neuronal web connections. The automobile is limited merely to the wiring and connections of its electrical system.

Fifth, the human body produces a vast variety of output functions. Simple output functions are force, torque, work, velocity, acceleration and momentum. Complex output functions are analyzing, synthesizing, inventing, and creating. The automobile has limited output functions such as acceleration, velocity, and momentum.

Sixth, the human body adapts to stress. It thrives and survives. It can even heal itself within certain limits. Some epithelial and connective tissue cells regenerate, bone collars form around fractures, wounds naturally heal, and many times the immune system defeats certain infectious diseases. The automobile does not repair itself like the human body. It slowly wears out. The mechanic may constantly replace parts and repair it to keep it running efficiently and safely. With proper care a car will last up to twenty years. The body should last on the average of 78 years and maybe into the eighties and nineties.

Seventh, the human body follows the law of use and disuse. If you use it, it develops; if you do not use it, it shrinks and loses function. There are four steps to this principle:
1. A load (stress) is the stimulus for biochemical activity.
2. The net result of this biochemical activity is catabolism, which is a destruction of cells and fluids.
3. This is followed by nutrition and rest.
4. The final result is anabolism, which is an overcompensating development of cells and fluids.

The reason the body develops from use is due to the overload principle, which states that in order for development to occur, a load above normal must be used in the activity. This is called the "rebound effect" or "overshoot phenomenon." Many of the body's tissues end up bigger and better. The overload principle is the essential foundation of successful physical training and rehabilitation. This principle is the reason that physical exercise causes humans to perform better at work or sports. It is the principle behind all the physical exercise and treatments used by physical therapists to return function to injured body parts. It is a beautiful, constructive response to active work. The automobile does not respond this way. Every time its parts are used, they wear out a little bit. They never rebound and get bigger and better. Only in animal bodies and especially the human body, do we find there the constructive effect of the overload principle. This truly is a magnificent response from a marvelous machine.

We can summarize the biological functions of the human machine in one sentence:

The human body is a carbon-based, chemically fueled, force fluid and air cooled, bipedal, communicative, photochromatic, binocular, cellularly self-replicating, self-diagnostic, self-repairing-tissue-wise, multidexterous, continuously adaptive, computer controlled (CPU chip not made by Intel), capable of short- and long-term memory with conceptual retrieval and integration, capable of precise decision-making and creativity, biodegradable exhaust system machine with a life expectancy (USA) of 75 years and a lifespan of 120 years. Truly, it is the Ultimate Machine!

Kaufmann, 1997

We know it takes an intelligent human designer (i.e., an inventive automotive engineer) to plan and construct our automobiles. Since designs require a creator, an unbiased observer would have great difficulty denying the rationality that the *Ultimate Designer* made the *Ultimate Machine,* the human body.

Chapter Fourteen

The Unseen Hand

The invisible hand that governs the universe with "perfect intentionality" has worked for the good of those who love him.

R. C. Sproul (1996, cover jacket)

Evolutionary Dogma

"Nothing in biology makes sense except in the light of evolution."

This is the title, thesis, and brazen pronouncement of the geneticist and committed evolutionist, Theodosius Dobzhansky (1973). It was Darwinian evolution, he maintained, that was the glue which held all areas of the biological sciences together. This article has become a watershed for the secular scientist which has been cited hundreds of times in magazines, journals and conference presentations that invoke evolution as fact. The article also proposes that evolution is integrated with everything in biology. Theodosius Dobzhansky was probably, next to Julian Huxley, the most influential evolutionist of the 20th century. What would cause Dobzhansky to make such a bold claim that nothing makes sense in science except in the light of evolution?

Molecular Evidence

In his article, Dobzhansky cites as evidence *cytochrome c*, the protein involved with the electron transport system. This important protein is indeed found throughout the plant and animal kingdoms. But does this mean that all these vastly different organisms have genetic roots from one common ancestor as Dobzhansky

maintained? It is more logical to argue that these biomolecules owe their existence to a common Designer, not a common ancestor. Since almost all cells utilize energy to survive (and therefore undergo cellular respiration), it is logical that the Creator would use the same molecule and efficient plan in cytochrome c to ensure the rapid transport of electrons and thus, energy.

The same logic would be used for the famous molecule of heredity, DNA. Dobzhansky thought that the universality of this information sequence in nearly all cells is evidence for a common ancestor of prokaryotes and eukaryotes. But it is just as logical to view the same plan being used to code for proteins in all living organisms. This is why the Mustang and Taurus have more in common than Ford cars and sailboats. It is the creation according to a common plan. Certainly as one looks at the code that is DNA, one would think of a Code-giver.

Microevolution in Fruit Flies

Dobzhansky also spoke of subtle changes within varieties of fruit flies in Hawaii as evidence of "evolution" occurring today. There is no question that throughout the years various selective pressures have given advantage to some populations of flies and not others. Creationists are certainly would agree to these minor changes called microevolution and to natural selection. The concept of natural selection, for which Charles Darwin is most famous, was actually described by a creation scientist named Edward Blyth 24 years before Darwin's publication of the idea (Parker, 1996). The question a student should ask is whether these minor variations of fly types lead to big changes (mega or macroevolution). The answer can be given as a clear no, using a number of evolutionary sources. They firmly state that one cannot extrapolate from micro to macroevolution. From beginning to end we start with, and end with, fruit flies, nothing more.

Embryological Evidences

Although dismissing Haeckel, Dobzhansky makes a new case for recapitulation theory in his article (1973). Dobzhansky gives some of the typical evidences for the recapitulation theory as discussed in Chapter 3. Students a generation ago were taught that "residual gill slits" appeared in the brachial region during embryonic development. These are now correctly called pharyngeal throat pouches. They have nothing to do with gills. Indeed, these structures turn into vitally important glands like the thymus and parathyroid.

On rare occasions, a child will be born with a taillike tumor on the back. Many people, doctors included, will unfortunately and unscientifically say that this structure is a residual tail, a throwback to our ancient tetrapod ancestors. Once again, good science says something different; it is more logical to conclude the function of the distal vertebrate are useful for pelvic muscle attachment. Sometimes, because of a mu-

tation, the folds continue to grow together, producing an outgrowth, that looks slightly like a tail. Such a tumor does not have vertebral bones inside, nor does it possess other structures we associate with a tail, like nerves and muscles.

The last word in this argument came in 1987 from a hospital researcher in London who was surprised to see that many textbooks still carried the deceptive and devious embryological wood cuttings of Ernst Haeckel of Germany. It was in the 1800s that Haeckel proposed the unscientific "biogenetic law" and doctored his illustrations to support this now-defunct idea. Even though the fraud was eventually exposed, millions of students continue to be fooled through the present day.

Is Evolution a Fact?

Macroevolution is unfortunately taught as a scientific (empirical) fact. Twenty-five years after Dobzhansky's pronouncement, this argument continues with evidence of change in antibiotic-resistant bacteria. Many newspaper and popular accounts of evolution frequently cite the growing medical threat of multidrug resistance in bacteria as proof for evolution.

Antibiotic-resistant bacteria are hardly evidence for macroevolution. Although creation scientists acknowledge changes of antibiotic sensitivity in bacteria where genes are traded and transferred among them, these bacteria are still the same kind. Recently, an expedition to the Arctic uncovered the bodies of three men who perished in an 1845 expedition. Samples of bacteria were taken from their intestines and it was found that some of the bacteria were indeed resistant to modern-day antibiotics. This is just as the creation scientist would predict. There have always been some populations of bacteria that have had genes conferring a resistance to antibiotics (Parker, 1996).

Plasmid transfer, conjugation, and transduction are most likely the mechanisms for changing antibiotic resistance among bacteria. These mechanisms, however, allow for horizontal change, but not vertical evolution in bacteria that are necessary for macroevolution to take place. Medical microbiologists, physicians, and hospital technologists are all interested in tracking the patterns of change in bacteria for practical reasons, being able to provide an effective antibiotic that will eliminate disease. Fortunately, they do not have to worry that the bacteria will evolve into a fungus or some other microbe. The bacteria are still bacteria and remain so, therefore, the practitioner can resolve the resistance factor with new and better antibiotics.

We can say therefore that science performed in the present (empirical science) would not support an extrapolation from very minor changes to macroevolution. Put another way, the only changes we see are horizontal (minor), certainly not large changes that would yield new kinds.

No macroevolution has occurred in the past, and none is occurring in the present. There is no reason to believe it happened except that the alternative is unthinkable to the secular mind. We would do well to investigate some of the alleged evolutionary evidences that Dobzhansky lists in his influential article. Several of the evidences actually line up quite well with the creation model.

Human Evolution

Java Man, Peking Man, Piltdown Man, Nutcracker Man, and *Ramapithecus* are all false starts on the trail from apelike creatures to man. Dr. Takahata (1995, p. 131) stated that, "However, there are not enough fossil records to answer when, where, and how *Homo sapiens* emerged." Sylvia Mader (1998, p. 353) stated the critical link between man and ape *"has not yet been found."* Evolutionary thinking has caused people to spend millions of dollars and thousands of man-hours looking in vain for links that will never be found. But the search continues for man's ancestor, because many still believe that *"Nothing in biology makes sense except in light of evolution."*

Imperfections

Richard Dawkins continues the theme of evolution as a fact when he states *"No serious biologist doubts the fact that evolution has happened, nor that all living creatures are cousins of one another"* (1996a, p. 287). He discusses the imperfections in nature, such as the aberrations of eyesight in humans. He turns to Darwinian evolution, the gradual tinkering of natural selection and chance mutation, to explain the evolution of the eye. He discusses how a neutral evolutionary force over time makes images to the eye sharper and sharper. He uses these examples of poor eyesight in man as proof of nature's tinkering with traits to produce a superior eyesight. Yet there are no examples for this type of evolution seen today, nor is there a history of eye transitions observed in the fossil record.

Defects and Disease

Imperfections of the eye (nearsightedness, farsightedness, and astigmatism) are merely a consequence of the fall of man. In the biblical account of Creation, man was not merely made a complex animal, but a creature made in the image of God. Shortly after his creation, man stained this image by disobedience to his Creator. Man's image became tarnished and any of his "imperfections," defects, and diseases are consequences of his fall from perfection. Yet, in spite of being marred by sin, we still observe many wonders in the human body. The intelligent design and many adaptations found in the human body are perfect testimonies to the Creator's wisdom and greatness (Silvius, 1997). Man continues to be sustained by the one *in whose hand is the soul of every living thing, and the breath of all mankind* (Job 12:10). We might summarize man's divine image as defaced, but not erased.

Recall that Leonardo da Vinci saw man as the pinnacle of Creation and a spectacular object of study in spite of his fall. Leonardo during the Renaissance period made the most anatomically comprehensive study of the human body. Although he did not view man as perfect, he saw humans as possessing the most wonderful and magnificent design in nature. He saw neither superfluous nor defective structures in man. Leonardo da Vinci looked for ways to describe this masterpiece of engineering, beauty, complexity, and symmetry. In his study of body proportions, he frequently made the analogy between man and machine.

Uniform Experience

Dr. Charles Thaxton argues for the "principle of uniform experience" to support the idea that there is an intelligent cause for life (Buell and Hearn, 1994; Davis and Kenyon, 1993). This principle may be illustrated from the "rock formation" that we discussed in chapter 2 (Mount Rushmore in South Dakota). Even though we did not see the construction process, we logically conclude from our experience that these stones were carved by workmen following a blueprint. This plan was devised in the mind of an architect ahead of its actual construction.

The idea starts on the drafting table of an architect. From there a blueprint is taken to the field to see if the plan is practical. Once adjustments are made construction begins on the monument (Figure 14.1). We recognize intelligence at Mount Rushmore, because of the *symmetry* in the stones as we see the heads of U.S. Presidents, the overall *interdependence* of the forms that compose the entire surface, and the obvious *purpose* revealed by the sculpture. We reject the notion that the shapes of these rocks are the product of wind and water erosion (Austin, 1994, p. 153). It is the action of human hands.

Likewise, according to our "experience," we can logically conclude that living systems with order, organization, complexity, symmetry, interdependence, and purpose are also products of intelligent design. From this principle of common or uniform experience, we may deduce that human DNA, like the Mt. Rushmore blueprint, is a predestined plan conceived by an Architect before its construction. As we consider the common experience of blood drawn from our body, we may also deduce that DNA provided genetic instructions for the manufacture of red blood cells. Each of these was devised by an intelligent cause.

Our best judgment tells us that informa-

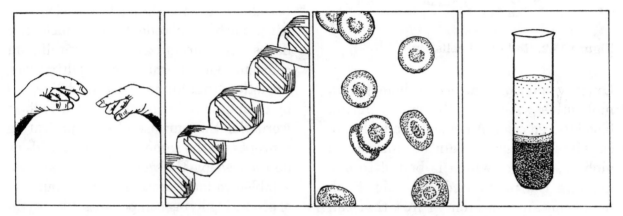

Figure 14.1. The Principle of Uniform Experience. The chain of logic, illustrated compares the formation of Mt. Rushmore (an artificial design) with formed components isolated from human blood (a living design). These eight illustrations apply this concept.

tion, a code, or a message takes an intelligent agent to generate it. We never see functional complexity derived from disorder, by chance, or by chaos. This is shown in Figure 14.1. We must logically conclude that patterns observed in the components of blood are by the action of God's unseen hand!

Red Blood Cell Design

Normal erythrocytes are nearly uniform in size, with diameters of 7.2 to 7.9 μm. The average red blood cell is about 2 μm thick at its thickest part. However, erythrocytes with much larger and much smaller sizes may be found in certain disease states. The shape of the normal eryth-

rocyte is a biconcave disc. This configuration allows for maximum surface contact of hemoglobin with the cell, thus greatly facilitating the exchange of blood gases. In fact, IBM did a study on the ideal shape of a solid object that would maximize oxygen diffusion. After running numerous simulations on an IBM mainframe computer, the biconcave disc, or "lifesaver," shape was determined optimal. Furthermore, this shape gives the red blood cell great flexibility and elasticity. This shape also allows it to be folded when it has to move through very narrow blood capillaries. The smooth, round edges reduce the amount of friction the cell may encounter in microcirculation.

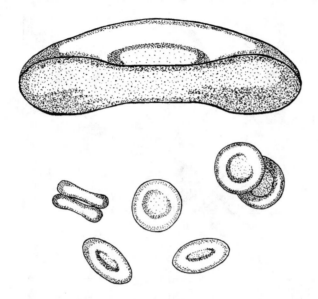

Figure 14.2. Red blood cells.

Oxygen molecules attach to hemoglobin molecules within erythrocytes, giving blood its red color. A hemoglobin molecule consists of four protein chains called globins, each of which is bound to one heme, a red pigmented molecule. Each heme contains an atom of iron that can combine with one molecule of oxygen. Thus the hemoglobin can transport up to four molecules of oxygen. Considering that each erythrocyte contains about 280 million hemoglobin molecules, a single erythrocyte can transport over a billion molecules of oxygen (Van de Graaff, 1998). It is within the lungs that oxygen molecules contained in inhaled air attach to the hemoglobin molecules and are transported via erythrocytes to trillions of body cells.

More Living Designs

The greatest biochemical discovery of the twentieth century was the unraveling of the chemical structure of chromosomes. The language of genetics is founded in the pairs of nucleotide bases (guanine, cytosine, adenine, and thymine plus uracil in the case of RNA). The bases are linked to form DNA and RNA. The sequence of bases in DNA and RNA is actually a code for amino acids, the building blocks for proteins, such as albumin, immunoglobulin, fibrinogen, and other proteins found in blood.

A remarkable analogy exists between human language and genetic language. By common agreement, numbers are the elements of computer language (such as Pascal), which are used by intelligent humans to code for letters. Similarly, the elements of the biochemical code (the four nucleotide bases) were organized by an intelligent cause to code for amino acids. Ultimately, the letters that humans use, as delineated by computer code, are assembled to form words for the purpose of communicating concepts. In the same way, amino acids as delineated by the genetic code, are assembled to form proteins for the purpose of producing living components, like erythrocytes, leukocytes, and platelets. The similarity between the two codes is great enough to support a general conclusion by our "principle of uniform experience." There must be an intelligent cause for both.

Intelligent humans devised the computer code in a fashion similar to the way an intelligent cause devised the genetic code. We note that human intelligence is external to the computer code, and we stipulate that an intelligent cause is external to the genetic code. Thus, the intelligence is not latent within the genetic language itself, as the pantheist might assume. The similarity in the DNA code exists because the Creator used a similar plan to fashion the diversity of life on earth. Likewise, the numbers in the computer code have no

Figure 14.3. Partial amino acid sequence for hemoglobin component.

phenylalanine-alanine-threonine-leucine-serine-
glutamine-leucine-histidine-cysteine-leucine

β-Chain for Human Hemoglobin

meaning, unless words and language exist external to the order. Similarly, the sequence of bases in DNA has no meaning without the amino acids (see Figure 14.3). The machinery external to the code builds proteins in red blood cells (Stryer, 1981).

Not only is the intelligence external to the genetic code, but the intelligence must have incredible sophistication. It can be concluded that, because man is just beginning to understand DNA structure, the origin of life was caused by a source with greater intelligence and manipulative ability than man. Yet even this is still an insufficient picture of the structure of this cause. DNA is only a small part of the simplest cell's complexity. Without the ability to reproduce itself, the DNA would eventually succumb to destruction. The process of replication, therefore, is essential to life. Without transcription and translation of DNA into cellular materials, the information is useless.

Thus, transcription and translation are necessary for life, as well. Upon consideration of these and the tremendous complexities of even the simplest cell, it becomes clear that the origin of the red blood cell must be assigned to a cause with incredible intelligence and manipulative ability.

Compared to the functional complexity of cells like erythrocytes, leukocytes, and platelets, man's most noteworthy technological achievements seem primitive. The discovery of the unique design and functional complexity of living organisms has caused many to doubt the naturalistic origin of life components in man. Logic demands that the languages of life (DNA, RNA, amino acids) be interpreted as indicators of God's work.

Sleuth Work

In analyzing the human body systems and their origin, you need to become a sleuth. Sleuths are detectives, trackers, or blood hounds who search for a criminal, or an elusive clue. The way to check the plausibility of an origins model is to consider circumstantial evidence. In a murder case with no eyewitnesses, detectives must rely on the strength of circumstantial evidence (Figure 14.4). Epidemiologists, or medical "sleuths", have done detective work such as discovering an epidemic in progress, defining circumstances under which a dis-

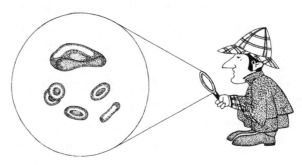

Figure 14.4. Nothing in biology makes sense except in light of the evidence.

ease spreads, looking for clues that might correlate a specific microbe with an infectious disease, studying the pattern of disease distribution, and applying statistics to find the probable cause of a given disease. Sleuths need keen observation, problem-solving, and critical thinking skills. Sleuths possess traits like that of Sherlock Holmes: patience, persistence, discipline and creativity. Sleuths must be open to new methods and ways of conducting their detective work, while maintaining their traditional trait of being a good "bloodhound."

As you examine new data in the organization of the human body, we encourage you to think critically, like a detective would. Be aware of the assumptions that go into each way of thinking from a design versus a blind (chance) watchmaker perspective. You will want to utilize a variety of methods and skills to understand the nature of human body origins. Should the disciplines of human anatomy and physiology be explained by intelligent design or by naturalistic descent? You must make an intelligent choice.

The Invisible Hand and His Word

The way of describing reality brings to mind a passage in the Gospel of John:

In the beginning was the Word, and the Word was with God, and the Word was God. The same was in the beginning with God. All things were made by him; and without him was not any thing made that was made. In him was life; and the life was the light of men.

John 1:1–4

This passage tells exactly how we would describe the creation of a literary work, a computer program, or a building. In the beginning was the concept and then there was the working out of that concept in the mind of the author or deity. Thereafter, the idea was recorded, or realized, in matter. The Word (information) is not reducible to matter; it precedes matter.

In contrast, the evolutionist views the beginning with only the material, such as particles and cells. From these cells came lower animals, then monkeys and finally man—literally, everything came from matter. Most Darwinists and reductionists of life view mankind as products of chaos. In regard to the human body, they leave no room for a Creator, much less the Word of the Creator (Johnson, 1997).

Conclusions

In conclusion, intelligent design in the human body may be observed in the purposeful arrangement of individual parts in the eleven body systems. The idea of design flows naturally from those observations and also from the Bible. The products of the human body appear not to be by chance and necessity, but rather they appear to be planned. New data from medical research studies corroborate the time-tested analogy that the human body and its functions are like a finely tuned machine. The Creator knew what completed and functional systems look like, then took steps to bring the systems into being. The human body at its most fundamental level and in its most critical components, (like a high tech machine) appears to be the product of intelligent activity.

Design is most evident when a number of separate, interacting components are ordered in such a way as to accomplish a function beyond the individual components. In general, the greater the specificity of the required, interacting components to produce a specific function, the greater the probability and confidence that its origin was by intelligent design.

In conclusion, in creating a new organism, as in building a new house, the blueprint comes first. We cannot build a fine house by tinkering with a tool shed and adding bits of material here and there. We have to begin by devising a plan that coordinates all the parts into an integrated whole. The same could be said of designing a new PC, fax machine, or Internet server. Today, designers of high tech operations spend months planning integrated, networked PCs, and online services before implemention. Intelligent design predicts the origin of the human body in a creative cause; it will be found in the blueprints, the plans, and the patterns devised by the Creator. In contrast, evolution models predict the origin of new organisms through chaotic, naturalistic, and material causes.

As examined, the processes in the human body are very complex in the eleven body systems and are evidences that a Designer is responsible for this marvelous body plan. Isaac Newton said *"In the absence of any other proof, the thumb alone would convince me of God's existence."* Today, more and more scientists are once again agreeing with Newton (e.g., Lester, Englin and Howe, 1998; Silvius, 1997). For over a century, very few scientists had a satisfactory, scientific answer to Darwin's theory, but an increasing number of professional biological scientists see Intelligent Design as a reasonable alternative to evolution (eg. Behe, 1996; Denton, 1998.)

The wonder of the human body is not new. In fact it was revered long ago. As we study the intricacies of the human body, we must conclude that its design was not only intelligent, but was wonderful because it is a product of a personal and skilled Creator. In fact, King David got it right when he penned these words over 3,000 years ago:

For Thou hast possessed my reins: Thou hast covered me in my mother's womb. I will praise thee, for I am fearfully and wonderfully made: marvelous are thy works, and that my soul knoweth right well. My substance was not hidden from thee, when I was made in secret, and curiously wrought in the lowest parts of the earth. Thine eyes saw my substance yet being unperfect. And in thy book all my members were written, which in continuance were fashioned when as yet there was none of them.

Psalm 139:13–16

Contemporary interpretation of blood clotting mechanisms by scientists is analogous to the interpretation of historical events made by various "wise men" during the day of Daniel and the reign of Belzhazzar. The Israelite prophet, Daniel, was a man who understood science (Daniel 1:4) and was gifted by God to interpret difficult enigmas. At a critical point in history, the Creator's hand became visible (Daniel 5) and wrote a message on the wall of Belzhazzar's palace for all to see. Although many of the king's choice scholars read the inscription, only Daniel could tell the meaning of the cryptic words. He was able to interpret the "handwriting on the wall" (Daniel 5:5). Daniel understood this writing because he trusted God to reveal the solution to this unsolved mystery.

Today the message is not one of written revelation, but one that is more subtle: the revelation of God's glory in His handiwork (Psalms 19:1), the human body. Sometimes molecular biologists encounter cryptic messages and unsolved mysteries in determining the mechanisms behind physiological patterns and pathological conditions. Sometimes dental researchers encounter anatomical anomalies, like the facial, sphenomandibularis muscle, and find its function puzzling. Biochemists study "biological messages" in the language of life, DNA, and in the intricacies of blood clotting and the immune response. In a sense, these microscopic observations are modern examples of "handwriting on the wall." As creation scientists, we must follow Daniel's example by asking God to illumine the mysteries, the themes, and the enigmas encountered in today's biological world. It is the Creator who gives us understanding of these patterns. Although the secular scientist works by his own "light" and ignores God in His thinking about the natural world, the creation scientist, in contrast, looks to the Creator's Word to illumine his mind for understanding. It is there he finds truth.

In Daniel's day, the message from the Hand was one of impending judgment. Today, with an understanding of scripture and a prudent study of biological "fingerprints", one can see a message in the physiological patterns. Today, the message is that the designs in the human body were crafted by an all-wise Maker and He cares for you, His creation. This same Creator is also Judge and His word says that we will all be accountable to Him someday (Romans 1:29). One can still see clear writing on the wall which tells us that someday we will meet our Maker who has been sending subtle messages to us all along. He is present in our world.

The Watchmaker is not blind as some contend, nor is the "glue" that holds all biology together, evolution. But rather it is the Hand of the Creator that holds all things together by His power (Colossians 1:16–17). Therefore, the best scheme that links (or "glues") all biology together is intelligent design. Although God may be invisible to our physical eyes, He provides us with sufficient "light" to "see" Him with spiritual eyes. The Bible is our divine revelation and the Father's world is our natural revelation. Both illumine our minds with evidence to understand the patterns of the human body. The Wisdom of God is revealed in the Living Word, Jesus Christ; in the written word, the Bible; and in nature. The natural "fingerprints" from His Invisible Hand are a testimony to the Creator's glory and His care for all life (Psalms 8:4; 19:1). The fact that this Creator also gives us a plan and purpose for life is our fortress, our shield, and our very great reward. This book only scratches the surface of human physiology, but we have presented adequate evidence to testify that the human body themes are in fact "fingerprints" from His Invisible Hand. Our prayer is that it may function as a stimulus for the reader to delve deeper into the matter.

References

Alcamo, I.E.1997. *Fundamentals of microbiology*, fifth edition. Benjamin Cummings, Redwood City, CA.

Access Excellence Editors. 1996. "Junk DNA" found at Access Excellence's What's News Site: http://www.gene. com/ae/WN/index.html.

ASA: American Scientific Affiliation—Committee of integrity in science education. 1986. *Teaching science in a climate of controversy*. American Scientific Affiliation Publications, Ispich, MA.

Austin, S.A., Editor. 1994. *Grand Canyon: Monument to catastrophe*. Institute for Creation Research, El Cajon, CA.

Ayoub, G. 1996. On the design of the vertebrate retina. *Origins and Design* 17(1):19–22.

Behe, M. J. 1996. *Darwin's black box: The biochemical challenge to evolution*. The Free Press, New York.

Bergman, J. and G. Howe. 1990. *Vestigial organs are fully functional*. Creation Research Society Books, St. Joseph, MO.

Berrill, N. J., and G. Karp. 1976. *Development*. McGraw-Hill Books. New York.

Bradley, W., and C. Thaxton. 1994. *The creation hypothesis*. Intervarsity Press, Downer's Grove IL.

Brand, P. and P. Yancey. 1993. *Pain: The gift nobody wants*. Harper Collins, New York.

Brand, P. and P. Yancey. 1984. *In his image*. Zondervan, Grand Rapids, MI.

Brand, P. and P. Yancey. 1980. *Fearfully and wonderfully made*. Zondervan, Grand Rapids, MI.

Buell, J. and V. Hearn, editors. 1994. *Darwinism: Science or philosophy?* Foundation for Thought and Ethics, Richardson, TX.

Campbell, N. 1996. *Biology*, fourth edition. Benjamin Cummings, Redwood City, CA.

Clark, M. E. 1976. *The amazing circulatory system: By chance or by creation*. ICR Research Technical Monograph Number 5. Creation-Life Publishers, San Diego, CA.

Cosgrove, M. 1987. *The amazing body human: God's design for personhood*. Baker Book House, Grand Rapids, MI.

Cox, D. L. 1992. Personal communication.

Darwin, C. 1979. *The origin of species* (Reprint of first edition). Avenue Books, New York.

Davis, P. and D. Kenyon. 1993. *Of pandas and people: The central question in biological origins*. Haughton, Dallas, TX.

Dawkins, R. 1996a. *The blind watchmaker*. W. W. Norton, New York.

Dawkins, R. 1996b. *Climbing mount improbable*. W. W. Norton, New York.

Denton, M. J. 1986. *Evolution: A theory in crisis*. Adler and Adler. Bethesda, MD,

Denton, M. J. 1998. *Nature's destiny*. The Free Press, New York.

Dixon, B. 1976. *The magnificent microbes*. Athenaeum, New York.

Dobzhansky, T. 1973. Nothing in biology makes sense except in the light of evolution. *The American Biology Teacher* 3:125–129.

Dunn, G. F., G. D. Hack, W. L. Robbins, and R. T. Koritzer. 1996. Anatomical observation of a craniomandibular muscle origination from the skull base: The sphenomandibularis. *Cranio*, 14(2):97–103.

Eakin, R. M. 1982. *Great scientists speak again*. University of California Press, Berkeley, CA.

Eidsmoe, J. 1987. *Christianity and the constitution*. Baker Book House, Grand Rapids, MI.

Evans, C. L. 1949. *Starling's principles of human physiology*, tenth edition. Lea and Febiger, Philadelphia.

Fisher, R. B. 1977. *Joseph Lister, 1827-1912*. MacDonald and Jane's Publishers, London.

Gazzaniga, M. S. 1998. The split brain revisited. *Scientific American* 279(1):50–55.

Gillen, A. L. and H. D. Mayor. 1995. Why do we keep catching the common cold? *The American Biology Teacher* 57(6):336–342.

Gillen, A. L. and R. P. Williams. 1994. Pasteurized milk as an ecological system for bacteria. In *Labs that work: The best of the how-to-do-its*. NABT Publications, Reston, VA.

Gish, D. T. 1995. *Evolution: The fossils still say no!* Institute for Creation Research, El Cajon, CA.

Glasser, R. J. 1976. *The body is the hero*. Random House, New York.

Gould, S. J. 1980. *The panda's thumb*. W. W. Norton, New York.

Gray, N. 1995. *Gray's anatomy,* fifteenth edition (Editors T. P. Pick and R. Howden, original edition, 1901). Barnes and Noble, New York.

Guyton, A. C. 1991. *Textbook of medical physiology*, eighth edition. W. B. Saunders, Philadelphia, PA.

Harvey, W. 1995 (reprint of 1628 edition). *De motu cordis: Anatomical exercises concerning the motion of the heart and blood in living creatures*. Dover Books, Mineola, NY.

Hole, J., D. Shier, J. Butler and R. Lewis 1996. *Hole's human anatomy and physiology*, seventh edition Wm. C. Brown, Dubuque, IA.

Hole, J. W. 1995. *Essentials of human anatomy and physiology*, fifth edition Wm. C. Brown. Dubuque, IA.

Jackson, W. 1993. *The human body: Accident or design?* Courier Publications, Stockton, CA.

Jaret, P. 1994. Viruses: On the edge of life. *National Geographic* 186(1):58–91.

Johnson, P. E. 1993. *Darwin on trial*, second edition, Intervarsity Press, Downer's Grove, IL.

Johnson, P. E. 1997. *Defeating Darwinism by opening minds*, Intervarsity Press, Downer's Grove, IL.

Katz, D. 1972. Tonsillectomy: Boom or boondoggle? *Detroit Free Press,* April 13, p. 1-C.

Kaufmann, D.A. 1997. Personal communication.

Kaufmann, D.A. 1995. Physiological evidences for creation. *Creation Research Society Quarterly* 31(4):239–247.

Kaufmann, D. A. 1994. Anatomical evidences for creation: Design in the human body. *Creation Research Society Quarterly* 31(1):35–41.

Kaufmann, D. A. 1981. Mechanical design in the human body. *Creation Research Society Quarterly* 11(2):155–158.

Kaufmann, D. A. 1980. Design in the human body. *Creation Research Society Quarterly* 11(2):91–94.

Leakey, R. E. 1982. *Human origins*. E. P. Dutton, New York.

Lester, L. P., D. L. Englin, and G. W. Howe. 1998. *Designs in the living world*. SimBioSys Publishers, Hull, GA.

Mader, S. S. 1998. *Biology*, sixth edition. McGraw-Hill, Boston. MA.

Marieb, E. N. 1994. *Essentials of human anatomy and physiology*, fourth edition. Benjamin Cummings, Redwood City, CA.

Marshall, G. 1996. An eye for creation. *Creation Magazine* 18(4) 19–21.

McMullen, E. T. 1998. *William Harvey and the use of purpose in the scientific revolution: Cosmos by chance or universe by design?* University Press of America, New York.

Menton, D. 1991. *Eye analogy*. On Creation Science Web site: http://emporium.turnpike.net/C/cs/.

Miller, K. 1994. Life's grand design. *Technology review* (Feb./Mar):25–32.

Miller, K. 1996. One of us? *Life Magazine*, November: 38–56.

Moore, K. L. 1980. *Clinically oriented anatomy*. Williams and Wilkins, Baltimore, MD.

Morris, H. M. 1995. *The defender's study Bible*. World, Grand Rapids, MI.

Morris, H. M. 1988. *Men of science, men of God*. Master Books, El Cajon, CA.

Morris, H. M. and J. D. Morris. 1996. *The modern creation trilogy*. Master Books, Green Forest, AR.

Morrison, T. F., F. D. Cornett, J. E. Tether, and P. Gratz. 1977. *Human physiology*. Holt, Rinehart and Winston, New York.

National Academy of Science Editors. 1998. *Teaching about evolution and the nature of science*. National Academy Press,

Washington, DC.

Nester, E. N., C. E. Roberts, N. N. Pearsall, D. G. Anderson, and M. T. Nester. 1998. *Microbiology: A human perspective*, second edition. WCB/ McGraw-Hill, Boston, MA.

Nuland, S. B. 1997. *The wisdom of the body*. Alfred Knopf, New York.

Nuland, S. B. (editor) 1998. *Exploring the human body: Incredible voyage*. National Geographic Society, Washington, D.C.

Paley, W. 1802. *Natural theology*. Lincoln-Rembrandt Publishers (Reprinted in 1997). Charlottesville, VA.

Paolella, P. 1998. *Introduction to molecular biology*. WCB/McGraw-Hill, Boston, MA.

Parker, G. 1996. *Creation: The facts of life*. Master Books, Colorado Springs, CO.

Parker, G., K. Graham, D. Shimmin, and G. Thompson. 1997. *God's living creation*. A Beka Books, Pensacola, FL.

Pennisi, E. 1997. Haeckel's embryos: Fraud rediscovered. *Science* 277:1435.

Silvius, J. E. 1997. *Biology: Principles and perspectives*. Kendall Hunt, Dubuque, IA.

SoRelle, Ruth. 1993. In the spirit of David: A key genetic finding. *The Houston Chronicle*, April 3, 1993.

Sproul, R. C. 1996. *The invisible hand*. Word, Dallas, TX.

Stevens, J. 1985. Quoted by David Menton. *The eye*. On Creation Science Web site:

http://emporium.turnpike.net/C/cs/.

Stryer, L. 1981. *Biochemistry*. W. H. Freeman, San Francisco, CA.

Takahata, N. 1995. A genetic perspective on the origin and history of humans. *Annual Review of Ecology and Systematics* 26:343–372.

Thibodeau, G. A. and K. T. Patton. 1997. *The human body in health and disease*, second edition. Mosby, St. Louis, MO.

Thomas, L. 1986. In Pode, R. M. (editor), *The incredible machine*. National Geographic Society, Washington, D.C.

Tortora, G. J. 1994. *Introduction to the human body: The essentials of anatomy and physiology*, third edition. Harper Collins, New York.

Tortora, G. J., B. R. Funke, and C. L. Case. 1997. *An introduction to microbiology*, fifth edition. Benjamin/ Cummings, CA.

Tortora, G. J. and S. R. Grabowski. 1996. *Principles of human anatomy and physiology*, eighth edition. Harper Collins, New York.

Tsiaras, A. (editor). 1997. A fantastic voyage through the human body. *Life Magazine,* February:59–81.

Van de Graff, K. M. 1998. *Human anatomy*, fifth edition. Wm. C. Brown, Dubuque, IA.

Van de Graff, K. M. and S. I. Fox. 1995. *Concepts of human anatomy and physiology*, fourth edition. Wm. C. Brown, Dubuque, IA.

Wells, J. and P. Nelson. 1997. Homology: A concept in crisis. *Origins and Design* 18(2):12–19.

Williams, R. P. and A. L. Gillen. 1991. Kitchen microbiology and microbe phobia. *The American Biology Teacher* 53(1):10–11.

Yancey, P. 1990. *Where is God when it hurts?* Zondervan, Grand Rapids, MI.

Appendix A

Glossary of Design Terms

This glossary clarifies word meanings that may have a specialized usage in the context of creation, anatomy, and physiology. For an extensive treatment of technical terms, you should consult a standard medical, biology, or English dictionary. Also useful may be an anatomy and physiology textbook, such as:

Van de Graff, K. M. and S. I. Fox (1995). *Concepts of human anatomy and physiology*, fourth edition. Wm. C. Brown, Dubuque, IA.

Hole, J., D. Shier, J. Butler, and R. Lewis (1996). *Hole's human anatomy and physiology*, seventh edition. Wm. C. Brown, Dubuque, IA.

Tortora, G. J. and S. R. Grabowski (1996). *Principles of human anatomy and physiology*, eighth edition. Harper Collins. New York.

Acclimatization—Reversible alteration (adaptation) of an individual's tolerance for environmental stress or conditions eg. altitude conditions. Physiological adjustment usually takes days, weeks or months.

Acidosis—A serious physiological condition in which hydrogen ions overwhelm the buffers of the blood, resulting in a life-threatening low pH.

Adaptation—Creationists use this term to refer to a short term change in response of a body system as a consequence of repeated or protracted stimulation; whereas, evolutionist s use this term to mean adjustment to environmental demand through the long term process of natural selection acting on the genotype.

Adaptational Package—Biological organisms are more than the sum of individual structures; their ability to function successfully is due to an entire "package of parts."

Aldosterone—Hormone secreted by the adrenal cortex that regulates the potassium and sodium balance of the blood.

Alkalosis—A condition in which there is an increase in alkaline substances in the blood, especially sodium bicarbonate.

All-or-none principle—In skeletal muscles, individual fibers contract to their fullest extent or not at all. In neurons, if a stimulus is strong enough to initiate an action potential, a nerve impulse is propagated along the entire neuron at a constant strength.

Analogy (analogous structure)—A body part similar in function to that of another organism, but only superficially similar in structure, at most. Such similarities are regarded, not as evidence of inheritance from a common ancestor, as in homology, but as evidence only of similar function.

Anatomy—The science of the structure of living organisms.

Articulation—The junction of two or more bones. It can also mean fit together.

Asepsis—freedom from infection; condition in a body that is free of pathogens, as in wounds or surgical incisions that are free from pathogenic bacteria, fungi, viruses, protozoans and parasites.

ATP (Adenosine triphosphate)—The universal energy-carrying molecule manufactured in all living cells as a means of capturing and storing energy. It consists of adenine with a sugar, ribose, to which are added, in linear array, three phosphate molecules.

Autonomic nervous system—that branch of the peripheral nervous system (PNS) that has control over the internal organs. Divided into the parasympathetic and sympathetic nervous systems.

Biochemistry—The chemistry of life; the study of the constituents of living organisms.

Biological clock—An innate physiological rhythm, such as metabolic rate.

B-lymphocyte—Specialized leukocytes processed in the bone marrow (or **bursa equivalent**); these cells give rise to plasma cells that produce antibodies.

Blood trauma—Wound that is internal, and the intrinsic pathway is triggered when blood comes into contact with the damaged endothelial cells. This initiates blood clotting, but this pathway takes more time than one induced by tissue trauma.

Bone marrow—It is the soft spongelike material in the cavities of bones. Its principle function is to manufacture erythrocytes, leukocytes and platelets (formed elements of blood).

Blind watchmaker—Richard Dawkins' metaphor for non-intelligent, purposeless, natural selection; expressing the idea that design characteristics seen in living things are not like the intelligent purpose and craftsmanship associated with watchmaking.

Blood—A modified connective tissue in which liquid plasma is separated from platelets, white and red cells.

Cascade—A series of chemical reactions, or events, that involve interdependent reactions

Cell—The basic unit of structure and function in living organisms.

Centrifugal force—The force tending to make rotating bodies move away from center of rotation; it is due to inertia.

Coagulation—Refers to blood clotting. The process by which platelets and soluble plasma proteins interact in a series of complex enzymatic reactions to finally convert fibrinogen into fibrin. It works on a positive feedback cycle.

Commissures—Bridges of neurons that connect the brain hemispheres.

Compound Traits—Interdependent parts working together in an organ or system.

Correlation (Complementarity) of structure and function—The close relationship that exists between a structure and its function. Structure determines functions.

Convergent evolution—Evolution of similar appearances in unrelated species.

Corpus callosum—Is the most massive of the brain commissures and typically the only one severed during surgery for epilepsy.

Creationism—The belief that the Divine Designer produced living kinds rapidly and separate from each other in relatively recent time.

Cyanosis—A blue pallor resulting from insufficient perfusion of body tissues with oxygen.

Darwinism—The theory that all living things descended from an original common ancestor through natural selection and random variation, without the aid of intelligence or nonmaterial forces.

Design—Is the purposeful arrangement of parts; a plan, a scheme, a project, or a purpose with intention or aim.

Dialysis—Refers to the process of separating molecules of different sizes using a semipermeable membrane.

Digestion—Hydrolysis. The process by which large food molecules are broken down into simpler ones, thus being able to be absorbed through the epithelium of the GI tract.

DNA—(deoxyribonucleic acid)—Nucleic acid found in all cells (except mature red blood cells); the genetic material that specifies protein synthesis in cells.

Electrolyte—A substance, such as a base, salt, or acid whose molecules, when in solution, separate into ions. Such a solution is capable of conducting an electric current.

Embryology—The biological science which deals with the development of an organism.

Extracellular—Outside of a cell or cells.

Erythrocytes—Red blood cells.

Extrinsic Pathway—This biochemical pathway of blood clotting occurs rapidly, within seconds if severe **tissue trauma** occurs. It is activated by tissue factor (thromboplastin) and this initiates the formation of prothrombinase. It has fewer steps than the intrinsic pathway, therefore it operates faster than the intrinsic pathway.

Emergent Properties—Properties found in living things that show the sum together is greater than the addition of the separate, independent parts.

Eosinophil—A granulated white blood cell characterized by large reddish staining cytoplasmic granules; associated with helminth infection and allergy.

Enteric Bacteria—Bacteria that reside in the intestinal tract. Most of these bacteria are not harmful to the human body and many are mutualistic with humans providing valuable vitamins and breaking down nutrients. Many of these beneficial bacteria are now called "Pro-Bacteria."

Evolution, (evolve)—Mere "change in living things over time"; but meaning also, descent with modification, and particular mechanisms to account for change such as natural selection and gene mutation.

Falsifiability—Falsifiable-test characteristic that can prove an idea, concept, theory wrong in an experiment. An idea that can be tested to be true or false.

Feedback inhibition—Biochemical mechanism by which a protein (usually an enzyme) is rendered inactive by combining with the product of an enzymatic pathway or reaction.

Fixity—The idea that species never change.

Functional information—The base sequence of DNA that codes for structures capable of functions.

"G" Force—Gravitational force; force that draws humans to the earth at a rate of 32 ft./sec. (1 G). Forces greater than 1 G exerted on humans may occur in moving vehicles (eg. jets, shuttles, etc.) where the intensity of this force affects the body's physiology. If the force with which he presses against the seat becomes 5 times his normal weight during pull-out from a dive, the force acting upon the seat is said to be 5 G.

Gene pool—The total genetic material in the population of a species at a given point in time.

Genome—The total DNA for a given organism.

Glucose—A monosaccharide, also known as "blood sugar."

Gradualism—The view that evolution occurred slowly over time, with transitional forms grading finely in a line of descent.

Great Physician—This is Jesus, the God-Man who heals men and women of physical disability, infectious disease, emotional and spiritual disorder. He gives dignity to those who are ashamed of their condition.

Hemocrit—A test used to determine the ratio of blood cells to plasma.

Hemophilia—Refers to several different hereditary deficiencies of coagulation.

Hemopoiesis—The production of red blood cells.

Hemostasis—Refers to the stoppage of bleeding and wound restoration.

Homeostasis—A state of body equilibrium, or the maintenance of an optimal, or stable, internal environment of the body.

Homology (homologous structure)—A body part with the same basic structure and embryonic origin as that of another organism. In the evolution model, it implies the common ancestry of the two, while in the creation model, it implies a common design.

Hyperventilation—State in which there is an increased amount of air entering the pulmonary alveoli, resulting in reduction of carbon dioxide in blood.

Hypothesis—In scientific method, 1. An educated guess, 2. A tentative explanation, or 3. A proposition that is to be confirmed by test.

Immunity—A specified defense against a disease.

Hypoxia—Reduction of oxygen supply to tissue below physiological levels despite adequate perfusion of the tissue by blood.

Information theory—A branch of applied mathematics which provides a measure of knowledge in any sequence of symbols.

Innervation—The nerve supply to a part.

Integration—The coordination of excitatory and inhibitory signals and processes received by the cell body and processed at the axon hillock.

Intelligent design (cause)—Any theory that attributes the action, function, or structure of an object to the creative mental capacities of a personal agent. In biology, the theory that biological organisms owe their origin to a preexistent intelligence.

Interdependent Parts—Body structures that are interdependent on each other. The body is one unit, though it is made up of many different types of cells and tissues. The body parts work together, cooperate and cannot exist without each other. This resulting condition of these parts working together is that the sum of the actions is greater than the addition of the separate, individual actions.

Irreducible Complexity— Interdependent parts of living (and non-living) things that cannot be reduced further without losing their intended function (e.g. mousetrap parts working together).

Intrinsic Pathway—This pathway of blood clotting is more complex than the extrinsic pathway and it operates more slowly, usually requiring several minutes. It may be activated by collagen, glass, and other chemicals. The intrinsic pathway is so named because the formation of prothrombinase starts by damaged endothelial cells exposed to collagen.

Jehovah Rapha—The God who heals. The God who makes whole and the One who restores a person's life of calamity to a life of dignity and usefulness.

Leukocytes—White blood cells.

Lipid—Compounds that do not mix with water including fats, phospholipids, and steroids.

Macroevolution—The model of large-scale changes, leading to new levels of complexity. A large change from one kind of plant or animal to a different kind (e.g., fish-to-amphibians). Also called vertical evolution. It is the idea that all living forms are related as branches of one ancestral "tree" or "bush."

Medulla oblongata—Section of the brain stem controlling blood pressure, heartbeat and breathing and other important functions.

Membrane—A thin layer of soft, pliable tissue.

Metabolism—The chemical changes that occur within the body.

Microevolution—Small-scale genetic changes, observable in organisms. A small change within a specific group (e.g., Darwin's finches; antibiotic-resistant bacteria). Also called horizontal evolution.

Monocytes—Largest of white blood cells. They are phagocytic and differentiate into tissue macrophages.

Morphology—The form or structure of an organism.

Mosaic—Patterns and designs showing contrasts among colors, pigments, structures, and/or organisms in their environment.

Mutation—A relatively permanent change in the DNA involving either a physical change in chromosome(s) or a biochemical change in the order or number of nucleotide bases in gene(s).

Mucociliary escalator—The mechanism where cilia of mucous membranes cells move particles along respiratory membranes and up the throat, where they are swallowed.

Naturalism (naturalistic)—The philosophy that everything in nature can be explained by the natural, physiochemical processes. (No supernatural explanation is permissible in this ideology.)

Natural selection—Process in nature where one genotype leaves more offspring for the next generation than other genotypes in that population. The explanation is that the elimination of less suited organisms and the preservation of more suited ones result from pressures within the environment or competition or both.

Negative Feedback—A primary mechanism for homeostasis, whereby a change in a physiological variable that is being monitored triggers a response that counteracts the initial fluctuation.

Neo-Darwinism—the concept of Darwinism with the addition of the concepts of modern Mendelian and population genetics.

Nerve impulse—Electrochemical change (action potential) traveling along a neuron.

Neuron—A nerve cell.

Nucleotide—The fundamental structural unit of a nucleic acid, or DNA, made up of a nitrogen carrying base, a sugar molecule, and a phosphate group.

Ontogeny—The development of an individual.

Order—A fixed or definite plan and system.

Organ—A structure of the body formed of two or more tissues and adapted to carry out a specific function.

Organ System—A group of organs that work together to perform a vital body function.

Organism—The living body that represents the sum total of all its organ systems working together to maintain life.

Organization—A unified coherent group or systemized whole.

Pathogen—Technical term for germ, or germ-causing microbe, such as virus, bacterium, protozoan that may be disease causing.

Phagocytosis—The process of engulfing other organisms, the particle containing vacuole fuses with a lysosome whose enzymes digest the food. An example of this is macrophages eating bacteria and other invaders of the body.

Phylogeny—The racial history of an animal, person, or plant.

PMN—Polymorphonuclear neutrophil, refers to white blood cells with a segmented nucleus (eg. neutrophil).

Plasma cell (plasmacyte)—A mature antibody-secreting B-cell found mainly in lymph nodes.

Positive Feedback—A physiological control mechanism in which a change in some variable triggers mechanisms that amplify the change.

Physiology—The science of the functioning of living organism.

Physiochemical Forces—Naturalistic view of how blind, mechanistic forces through chance and natural selection over millions of years may explain the irreducible complexity of anatomical structures and physiological functions of the human body.

Pseudostratified columnar epithelium—Tissue that has cells (like bricks on end) with relatively large cytoplasmic volumes.

Punctuated equilibrium—The theory that speciation occurs relatively quickly, when spurts of rapid genetic change "punctuate" the "equilibrium" of primarily constant morphology (stasis) over geologic time.

Pyemia—Pus-forming infection.

Recapitulation—The idea that development in the embryo or fetus replays evolution. Ernst Haeckel first stated this idea in his so called Biogenetic Law: This is sometimes stated as, "Ontogeny recapitulates phylogeny."

Reticular formation—A complex brain network that arouses the cerebrum. It is also a site for the integration of various brain signals.

Retinal Pigment Epithelium (RPE)—A structure in the posterior portion of the eye that maintains the photoreceptor structure and function and is necessary for optimal vision. It provides nutrients for the retinal cells and enhances new cell growth in the photoreceptor region of the eye.

Selection pressure—The intensity with which an environment tends to eliminate an organism, and thus its genes, or to give it an adaptive advantage.

Selective advantage—A genetic advantage of one organism over its competitors that causes it to be favored in survival and reproduction rates over time.

Septicemia—Blood infection by bacteria, such as *Streptococcus pyogenes*.

Simple columnar epithelium—This tissue has cells with relatively large cytoplasmic volumes, and is often located where secretion or active absorption of substances are important functions. Their cells are shaped like bricks on end.

Simple cuboidal epithelium—This tissue is specialized for secretion and makes up the epithelia of thyroid gland shown. These cells are like sugar cubes in shape and secrete hormones into the bloodstream.

Simple squamous epithelium—This tissue is relatively leaky tissue. These cells (shaped like floor tiles) are specialized for exchange of materials by diffusion.

Small intestine—Long, tube like chamber of the digestive tract between the large intestine and stomach; divided into 3 areas.

Speciation—The development of a new species by the subdivision of an ancestral population.

Synapse—Region between two nerve cells where the nerve impulse is transmitted from one to the other, usually from axon to dendrite.

Sphenomandibularis—A new muscle that was recently discovered by cutting a cadaver from an unconventional angle and exposing an unfamiliar muscle connecting the mandible to the sphenoid bone behind the base of the eye socket. It is thought to stabilize the jaw during chewing rather than actually moving the jaw.

Systemic pressure—Arterial blood pressure during the systolic phase of the cardiac cycle.

Stasis—The constant morphology of a species over a long period of geologic time.

Synergism—Usually refers to properties found in non-living things (especially chemicals) that show the sum together is greater than the addition of the separate, independent parts. (Occasionally refers to properties of organisms.)

Taxon (plural, **taxa**)—A category of biological classification at any level.

Tight junction—Junction between epithelial cells where the membranes are in close contact, with no intervening spaces. Helps to maintain boundaries in cells.

Tissue—A group of functionally similar cells forming a distinct structure.

Tissue trauma—This trauma is caused by puncture of skin, or severe wound that is internal. It is so named because the formation of prothrombinase is initiated by tissue factor (TF), leading to the extrinsic pathway. Response is very rapid.

T-lymphocyte—Specialized leukocytes processed in the thymus gland; these lymphocytes provide the body with specific cell-mediated immunity.

Transitional—Refers to an organism that holds, in common with other organisms presumed to be its ancestor and descendent, particular features or structures of body form not held in common by the other organisms.

Urinary system—The organs concerned with the formation, concentration and elimination of urine. (synonym: excretory system)

Vestigial structure (organ)—A body part that has no function, but which is presumed to have been useful in ancestral species.

Vitamin K—Fat soluble vitamin, K_1 is found in foods and K_2 is produced by *E. coli*. Needed for the clotting of blood.

Appendix B

Types of Blood Cells

Type of Cell	Number per Liter	Percent
RBCs	5.00×10^{12}	94.1800%
WBCs	8.74×10^{9}	0.3200%
Granulocytes	5.24×10^{9}	0.1000%
Neutrophils	5.00×10^{9}	0.0942%
Eosinophils	2.00×10^{8}	0.0038%
Basophils	4.00×10^{7}	0.0008%
Monocytes	4.00×10^{8}	0.0075%
Lymphocytes	3.00×10^{9}	0.0565%
B cell	2.00×10^{9}	0.0377%
T cell	1.00×10^{9}	0.0188%
Natural Killer (NK)	1.00×10^{8}	0.0019%
Platelets	3.00×10^{11}	5.6500%
Total	$\mathbf{5.31 \times 10^{12}}$	**100.0000%**

Modified from Tortora and Grabowski, 1996, p. 565.

About the Authors

Dr. Alan L. Gillen wrote *The Human Body: An Intelligent Design*™. Dr. Gillen is an experienced biology instructor at the high school and college level. He has experience teaching general biology, microbiology, parasitology, and anatomy and physiology for 20 years. Presently, he teaches at Pensacola Christian College in Florida. He has authored (or co-authored) over 20 papers in biology, science education, and medical issues. In addition to writing this book on human biology, he has authored a companion book, *Body by Design: The Anatomy and Physiology of the Human Body*. This book provides anatomical evidence for creation. He has been active in bringing a biblical perspective in biology and medicine in the classroom and local churches for the last eight years. He has also contributed to *Creation Matters,* a CRS newsletter. One of his other favorite topics to speak and write about is *Listening to the Great Physician*. He has written a booklet by this title, and it is available upon request. His main professional interests are presenting human biology and medicine from a biblical perspective.

Dr. Alan L. Gillen

Mr. Frank J. Sherwin

Mr. Frank J. Sherwin is a frequent speaker on creation/evolution issues at church conferences and an experienced biology instructor at college level. Presently, Mr. Frank Sherwin is an assistant professor at the Institute for Creation Research, where he writes creation based curriculum and speaks at Back-to-Genesis seminars around the country. In addition, he prepares and speaks on the radio log *Science, Scripture and Salvation*. Previously, he taught general biology, parasitology, microbiology and anatomy and physiology at Pensacola Christian College for over nine years. His interests include: parasitology, jogging, and Shakespeare.

Mr. Alan C. Knowles

Mr. Alan C. Knowles is a popular anatomy and physiology instructor among nursing students at Pensacola Christian College. Mr. Knowles' main research interest is the nervous system and how it affects animal behavior. He has twenty-eight years of experience teaching science from a biblical perspective on both the high school and college levels. He joined the faculty of Pensacola Christian College in 1988. He teaches general biology, anatomy and physiology, and chemistry. His interests include biochemistry, computer science, and biological illustration. Mr. Knowles is responsible for some of the illustrations and computer graphics in this book.

Dr. George F. Howe was the editor of this book. He is a Creation Research Society board member and former editor of the *Creation Research Society Quarterly*. Dr. Howe has published numerous articles in the *Creation Research Society Quarterly* and has coauthored two popular biology books, *Designs in the Living World* and *Vestigial Organs are Fully Functional*. His research interests include plants living in the desert.

Index

Credits

Photographs

Cover—Color-enhanced SEM of blood clot, *Foundation for Thought and Ethics*

Figure 11.2—David Vetter photo, Baylor College of Medicine

Figure 13.1—David L. Cox cell biology pictures and Alan L. Gillen

Line Art

Figure 1.1—Mike Daily

Figure 2.1—Miriam Rodriguez

Figure 3.1—Miriam Rodriguez

Figure 4.1—Miriam Rodriguez

Figure 4.2—Miriam Rodriguez

Figure 4.3—Miriam Rodriguez

Figure 4.4—Miriam Rodriguez

Figure 5.1—Alan L. Gillen, modified from Hole, Shier, Butler, and Lewis (1996)

Figure 5.2—Alan L. Gillen, modified from Hole, Shier, Butler, and Lewis (1996)

Figure 6.1—Miriam Rodriguez

Figure 6.2—Miriam Rodriguez

Figure 6.3—Miriam Rodriguez, David L. Cox *E. coli* SEM, and Alan L. Gillen

Figure 6.4—Miriam Rodriguez

Figure 8.1—Alan L. Gillen, modified clip art

Figure 8.2—Miriam Rodriguez

Figure 11.1—Miriam Rodriguez

Figure 11.3—Miriam Rodriguez

Figure 11.4—Miriam Rodriguez

Figure 12.1—Miriam Rodriguez and Alan C. Knowles

Figure 12.2—Alan C. Knowles

Figure 13.1—Alan L. Gillen, modified clip art

Figure 14.1—Miriam Rodriguez, modified from ICR art file

Figure 14.2—Miriam Rodriguez

Figure 14.4—Alan L. Gillen, modified from *MacSleuth* (Gillen and Mayor, 1996)

Graphs
Figure 12.3—Data from Brand and Yancey, 1993

Figure 12.4—Data from Brand and Yancey, 1993

Figure 13.3—Data from Guyton, 1991

Charts
Figure 5.3—Acid base balance in the human body, Alan C. Knowles, modified from Hole, Shier, Butler, and Lewis (1996)

Figure 7.1—Outline of blood clotting, Alan C. Knowles, modified from Davis and Kenyon (1993), with permission.

All other charts, graphs, and figures done solely by the authors.